MATH PLANS

MATH PLANS

Lessons from the Field

*Patricia Marshall and Student Teachers
from the School of Education, CSU Sacramento
Illustrated by Ginger Herte*

Allyn and Bacon
Boston   London   Toronto   Sydney   Tokyo   Singapore

Copyright © 1997 by Allyn and Bacon
A Viacom Company
160 Gould Street
Needham Heights, Massachusetts 02194

Internet: www.abacon.com
America Online: keyword: College Online

All rights reserved. No part of the material protected by this copyright may be reproduced or utilized in any form or by any means, electronic or mechanical, including photocopying, recording, or by any information storage and retrieval system, without the written permission from the copyright owner.

ISBN 0-205-16270-3

Printed in the United States of America

10 9 8 7 6 5 4 3 2 1   00 99 98 97 96

This book is dedicated to math students at all levels.

Contents

Introduction   16
1 PREPARING TO TEACH
Children in K - 8   18
Developing Mathematical Concepts   19
Manipulatives   20
Lesson Plans   21
Speaking to 30+ Students   28
The Strands   36
    Geometry   36
    Measurement   42
    Probability and Statistics   44
    Patterns and Functions   47
    Logic   49
    Number   52

2 GEOMETRY
Things that Stick Out in the World *by Barbara F. Castrillo*   56
Bubble Magic *by Barbara F. Castrillo*   60
A Lesson Design in Geometry *by Linda C. Bolin*   62
Shapes with Felt and Yarn *by Elizabeth Castillo*   65
The Surprising Circle *by Baljinder Dhillon-Shergil*   67
Alphabet Symmetry *by Julia Serences*   69
Open and Closed Paths *by Karen Harrington*   72
Sorting Shapes *by Arleen R. Brienza*   74
Triangle Designs *by Pat Marshall*   77
Similarity and Congruence *by Anita Laca, Cheryl Walter & Jim Cordellos* 80
Rotational Symmetry *by Ortencia A. Wiley*   83
Polyhedrons *by Suzanne T. Tallman*   86
Battleship Game *by Yvonne B. Manley*   90
Circles *by Joy Peacock*   92
Projective Geometry *by Margaret-Ann Vansoest*   94
Triangle Inequality and Triangle Nests *by Rebecca B. Fuentes*   97
Fun and Folklore with Tangrams *by Sara Fries*   99
Tesselations *by Patricia Welty*   103
Geometric Quilt Bulletin Board *by Judith Nelson-Ullery*   107

3 MEASUREMENT
Grab Bag Comparisons *by Michele Denton Hanel*   110
Heavier and Lighter *by Kathleen A. Sutphen*   113
Time Order *by Kathy Mather Patschke*   115
Paperclip Measuring *by James Gillespie*   120
Ready, Set, Measure! *by Nancy A. Noma*   123
Making a Meter Tape *by Priscilla Huff*   126

Stringing 'Em Along by *Melissa K. Kassis*	128
Round Things by *Susan Macaluso*	131
Big Foot by *Pat Marshall*	133
Slicing Rectangles to Find Area by *Jennifer Olds, Jennifer Haskins and Dani Doiron*	136

4 PROBABILITY AND STATISTICS

Probability Detectives by *Marion Steed*	139
Graphing My Favorite by *Susie Beiersdorfer*	141
Flavor Favorites by *Janet Hooper*	145
Our Pets by *Cheryl Kreuzer*	149
Graphing Eye Color by *Kathy A. Sindel*	152
Coin Tossing Activity by *Christy Dalton*	157
The Jolly Rancher Draw by *Cathy Evans*	159
Seeing Red and Feeling Blue by *Julie Hayden Black*	161
Beans! Beans! Beans! by *Jacqueline A. Barnett*	163
What's in the Bag? by *Tim Cady*	166
Probability by *Brandi Zarzana*	168
Handedness by *Patricia Marshall*	171
Fair and Unfair Spinners by *Maria Lamirande*	174
Math For Health by *Anna Brandt Mayfield*	179
Beat the Teacher by *Julie Richards*	182
Fun With M&Ms by *Andrea Hurlbut*	185
Roll of the Dice by *Kelly R. Timpson*	187

5 PATTERNS AND FUNCTIONS

Show You Know by *Suzanne Lilliedoll*	190
Stamping Out Patterns by *Patti Bartholomew*	192
Staircase to the Clouds by *Paulette Johnston*	195
Leftovers by *Elizabeth Anne Brothers*	197
The Magic Box by *Pamela Howe & Christie Sonmez*	200
Cubes and Ways by *Debra R. Caldwell*	204
Looking for Patterns by *Gina Trzaska and Krisy Nettleton*	208
Patterns in Multiplication Facts by *Theresa J. Olmscheid*	211
Powers by *Patricia Marshall*	216
Can You See It In A Name? by *Cheryl L. Noack*	219
Patterns and Functions by *Gary T. Winegar*	221
Side-Stacked Squares by *Mary Charlesworth-Eggers*	224
I.M. Square by *Mary Simpson*	227
Chips and Dips by *Nancy E. Casper*	230
Handshake Problem by *Sharon L. Hagen*	233
Basic Functions by *James Hirleman*	238
Function Machines by *Penny Hall*	242

6 LOGIC

Guess the Secret Rule by *Carolyn DeVere*	246
Change One by *Carolyn Hoffman & Karen Szakacs*	248
Guess Two by *Marsha Heckert*	250
Dead Rat by *Pat Marshall*	252
Guess My Number by *Kim Leclaire*	254

Color In by Nancy J. Woods	256
Four in a Row by Deborah A. Allen	258
Logical Thinking by Felice Dinsfriend	260
Nuts by Pat Marshall	262
Poison by Ginni Loscuito	264
Matrix Logic by Jill Hohenshelt-Veach	266
Logical Algebra by Pat Marshall	271
King Arthur's Problem by Deborah M. Engelhart	274
The Problem of the 21 Water Casks by Jim Casey	276

7 NUMBER

Beginning Number Concepts

Unifix Trains of 1-6 by Jennifer Moldrem	281

Place Value

Counting by Tens with Kip by Jim Cordellos	284
Flats, Longs & Units by Kari Pfeiffer	287
Clear The Board by Hallie Atkinson	290
$1 000 Bill Game by Kelli Burns	292

Basic Facts

Sums of Seven by Darcy Cooper	300
Help for Walter by Michelle Richter and Tami King	302
Subtracting Numbers Less Than Ten by Monika Kuester	306
"Counting Up" with the Cover-Up Game by Loree Saberin	309
Colliding Cubes by Alan Hill	312
Missing Addends in <u>The Little Red Hen</u> by Carrie Callett	315
Tally Marks by Patricia Marshall	317
Spatial Multiplication with Color Tiles by Heidi Dettwiller and Caroline Miller	319
LCM by Dani Doiron	321
Dividing a Number by Itself by Paula Fallis	325
Word Problems in Multiplication and Division by Suzanne Blakeney	327

Algorithms

Explaining + and - Situations with Base 10 Materials by Pat Marshall	330
Using the Hundreds Chart to Add & Subtract by Pat Marshall	335
Showing Partial Products Using Rectangles by Pat Marshall	338
Dividing Money from a Cash Drawer by Pat Marshall	341
Hundreds of Dollars in the Cash Drawer by Pat Marshall	348

Fractions and Decimals

Fraction Freddy by Karen Wong	352
The Doorbell Mystery by Arlis Groves	355
Equivalent Fractions by Karen Wong	360
Geoboard Fractions by Carolyn Hoffman	364
Problem Solving with Fractions by Jennifer Emery	367
Fraction Strips by Karen Szakacs	371
It All Adds Up! by Christie J. Sonmez	373
"Tenths" Using Base 10 Blocks by Susan Frost	379
Multiplying Decimals by Jennifer Emery	382

Rates, Ratios and Percents

Fractions as Rates by Darcy Cooper	387
Ratios - Cubes to Tiles by Holly Whalen	390
Introducing Percents by Michael Knofler	393
Fun With Percents by Lori Souza	397

Appendices
Blackline Masters

Tangram	403
Spinners	404
Geoboard Template	405
Geoboard Recording Paper	406
Centimeter Dot Paper	407
Isometric Dot Paper	408
Inch Grid Paper	409
Half-Inch Grid Paper	410
Quarter-Inch Grid Paper	411
Rulers	412
Pattern for Base-10 Materials (inches)	413
Base-10 Mat	414
Hundreds Square	415
Hundreds Charts	416
Fraction Strips	417
Fraction Models	418
A Self-Critique of an Exploring Lesson in a Classroom That Has Never Used Manipulatives	419
Group Processing	422
Trouble Shooting an Exploring Lesson	423
A Generic Rubric for Scoring Open-ended Questions for Assessment	424

Bibliography 426

Index 432

FOREWORD

We at California State University, Sacramento are proud of the high quality work that our student teachers carry out as they learn to be teachers. The multiple subject student teachers who wrote and taught the lessons in this book provide children with their first experiences in all of the academic subjects and, as such, impart not only concepts and skills but also beliefs and attitudes toward every subject they teach. These lessons follow the spirit of the *Model Curriculum Guide*. They show planning in mathematics, a particularly sensitive part of the school curriculum, as it should be. The lessons are well prepared, the math is sound, the children are allowed to be themselves, and the teacher has the opportunity to grow professionally as she observes real children in the process of learning math.

Families need all the help they can get in these troubled times and thus the responsibilities of professional teachers have increased. These lessons show how groupwork can be carried out in the classroom. In emphasizing the practical implementation of groupwork, this book shows a way for teachers to respond to the needs of children.

This book will strengthen the math fieldwork component of student teaching for multiple subject teachers and will be a valuable resource to practicing teachers.

                                        Donald R. Gerth, President
                                        California State University, Sacramento
                                        May1995

ACKNOWLEDGMENTS

I would like to acknowledge the following student teachers for their contributions to this book:

Deborah A. Allen, Kevin Ashworth, Hallie Atkinson, Jacqueline A. Barnett, Patti Bartholomew, Susan Beiersdorfer, Julie Hayden Black, Suzanne Blakeney, Linda C. Bolin, Arleen R. Brienza, Elizabeth Anne Brothers, Elizabeth Broyles, Kelli Sue Burns, Tim Cady, Debra R. Caldwell, Donna M. Carrillo, Nancy E. Casper, Jim Casey, Elizabeth Castillo, Barbara F. Castrillo, Mary Charlesworth-Eggers, Kathleen Coates, Carrie Catlett, Rebecca Cocilova, Darcy Cooper, Jim Cordellos, Lisa Cox, Christy Dalton, John Dennison, Heidi Dettwiller, Carolyn E. DeVere, Felice Dinsfriend, Dani Doiron, Sara Douglas, Cheryle Ehrlich, Patsy Enderle, Deborah M. Engelhart, Katherine Estrada, Cathy Evans, Paula Fallis, Sara Fries, Susan A. Frost, Rebecca B. Fuentes, Rory M. Gay, Trina Gillen, James Gillespie, Christy Goldthwaite, Arlis Groves, Sharon L. Hagen, Penelope Ann Hall, Michele Denton Hanel, Karen Harrington, Jennifer Haskins, Kim Higgins, Alan Hill, James A. Hirleman, Jill Hohenshelt-Veach, Janet Hooper, Pamela Howe, Carolyn Hoffman, Priscilla Huff,, Andrea Hurlbut, Paulette L. Johnston, Melissa K. Kassis, Jeanne Keller, Tami King, Michael Knofler, Cheryl Kreuzer, Monika Kuester, Anita Laca, Maria Lamirande, Kim Leclaire, Suzanne Lilliedoll, Mary Longacre, Ginni Loscuito, Leanne Louch, Susan Macaluso, Rondine Mangrum, Yvonne B. Manley, Darlene Markey, Alma Jolene Matson, Anna Brandt Mayfield, Jennifer Moldrem, Caroline Miller, Judith Nelson-Ullery, Kristine Nettleton, Cheryl L. Noack, Nancy A. Noma, Jennifer Olds, Kathleen Mather Patschke, Theresa J. Olmscheid, Joy Peacock, Kari Pfeiffer, Linda J. Poisner, Julie Richards, Stacy Savoie, Loree Saberin, Julia Serences, Baljinder Shergill-Dillon, Mary Simpson, Kathy A. Sindel, Christie J. Sonmez, Lori Souza, Marion Steed, Kathleen A. Sutphen, Karen L. Szakacs, Suzanne T. Tallman, Kelly Timpson, Gina Trzaska, Margaret-Ann Vansoest, Cheryl Walter, Patricia Welty, Ortencia A. Wiley, Gary T. Winegar, Michelle Williams, Karen Wong, Nancy J. Woods, Brandi Zarzana

I would also like to acknowledge the following people for their help in this project:

Theresa McGinn, who typed and formatted the original version of this book; Dr. Larry Hannah and Mike Doyle for consulting on the MacIntosh; Dr. Steve Gregorich for his encouragement during his time as Dean; Dr. Patricia Roberts for her moral support; Phyllis Stenman and Suzy Lunstead for their gracious help in many ways; Lynda Nakamura for assisting in the final book preparation; my family, R. Stuart and Stu Marshall, for their encouragement and endurance; and Scarlet Maurin for putting together the book in its final form. I also wish to express my appreciation to the cooperating teachers who provided time and support for the professional development of these and other beginning teachers.

Time to complete the original **Math Plans** was provided by the Semester Leave with Pay Program at California State University, Sacramento. I would like to thank the individuals involved in granting this time for their vote of confidence.

ABOUT THE AUTHORS OF THE BOOK

The author and editor of this book is an associate professor at California State University, Sacramento where she has taught and supervised student teachers since 1983. She has a doctorate in mathematics education from Stanford University and worked for seven years previous to that as a multiple subject teacher specializing in math.

The authors of most of the individual lessons were first semester student teachers in the San Juan, Rio Linda and Multicultural Student Teaching Centers in the School of Education at CSUS. They came from a wide variety of backgrounds and were preparing to become multiple subject teachers in the grades from kindergarten through grade eight.

The illustrator is the GATE coordinator and fifth-grade teacher at Fairbanks School in the Del Paso Heights School District in Sacramento. She is also a graduate of the multiple subject credential program in the School of Education at CSUS.

Preparing to Teach

CHAPTER 1
Introduction

Children in K - 8
Mathematical Concepts
Using Manipulatives
Lesson Plans
Speaking to 30+ Students
The Strands
    Geometry
    Measurement
    Probability and Statistics
    Patterns and Functions
    Logic
    Number

INTRODUCTION

This book is for pre-service and in-service teachers who are trying new forms of teaching in their mathematics classes. It contains lesson plans on the topics of geometry, measurement, probability and statistics, patterns and functions, logic and number that were written and taught by beginning teachers. The ideas for these lessons come from the resources available to these beginning teachers: their cooperating teachers and professors, current books and periodicals, videos and workshops.

Two forms of math teaching, direct instruction and exploring, provide the lesson formats for all of the lessons in this book. These two lesson forms convey two different messages about math to students and the teachers who teach them. Direct instruction conveys the message that math is a body of knowledge that teachers already know and students need to know since, in this form of teaching, students learn by listening to and applying the teacher's instruction. An exploring format, on the other hand, conveys the message that math is an exploration of real things since, in this form of teaching, students learn by participating in and reflecting on rich activities chosen by the teacher. Both are valid forms of teaching. Whereas the format for direct instruction has been around for a while and is more familiar to many practicing teachers, the format for exploring lessons is newer and is now being developed as part of the repertoire for professional teachers. The majority of lessons in this book are exploring lessons.

As a guide to what primary and intermediate grade children can understand, we used the Essential Understandings shown in the *Mathematics Model Curriculum Guide* for the State of California. The *Model Curriculum Guide* lists several understandings for each mathematical strand and emphasizes the place of these understandings in the curriculum:

> These essential understandings bind together rather than precede all those specific concepts and skills which have traditionally been taught. They are the broad global ideas that expand or build, flower or evolve - that grow more complete and complex over time. (*Mathematics Model Curriculum Guide, Kindergarten Through Grade Eight,* 1987, p. 15)

The essential understandings for each mathematical strand are shown in the second half of chapter one. As you read the lessons in each of the following chapters, notice how each lesson embodies one or more of these essential understandings.

Preparing to Teach 17

The appendix contains a self-critique of an exploring lesson written by a student teacher in a classroom that has never used manipulatives. This self-critique provides a realistic glimpse of what to expect the first time you teach in a class that has not yet developed a groupwork infrastructure. The appendix also contains three post-teaching tools for teachers to use to improve and refine the "rough," first dry-run of an exploring lesson. The first tool, "Group Processing," is a worksheet to use with students to help them reflect on how to make the groups work better. "Trouble Shooting an Exploring Lesson" is a list of common pitfalls that can turn a wonderful exploring lesson into "The Lesson From Hell." This tool is to be used after teaching a lesson in order to sort out what went well and what needs refinement. The third tool, "A Generic Rubric for Scoring Open-ended Questions for Assessment," is a 0 - 6 scale that can be used to set up a scoring code for the assessments at the end of each lesson.

There is much to be learned about the teaching of math. The lessons described in this book should take students and their teachers to deeper levels of reasoning through experiences that impart understanding.

Patricia Marshall
Sacramento, California
April 1995

PREPARING TO TEACH

As you prepare to teach groups of children at various age levels, it is helpful to know what to expect of them. This chapter will give you an overview of planning and teaching active lessons to children in the grades from kindergarten to grade 8.

Children in K - 8

If your experiences prior to beginning teaching have been only with very young children or only with older children you will need to take some time to get a feeling for how an unfamiliar age group approaches math. One thing that remains constant throughout the grades is the use of perception as a major mode of learning new mathematical concepts. This means that before children can understand the meaning of new mathematical words or symbols, they must have experience with them in some way through seeing, touching and hearing. In conjunction with perception, the use of imagination and story-telling are also appropriate modes for teaching math in these grades. Based on observations and discussions with teachers, we offer these brief characterizations of the grades to beginning teachers:

K and 1: To children of this age level, the realm of possibility is limitless and make-believe occupies a substantial part of their thinking. The line between fact and fiction, reality and imagination is not yet set. Magic is still a possibility for them. Forming objects (using mud, clay, Plasticine, or beeswax, for example), as well as counting, sorting, and ordering objects are necessary and important parts of their mathematical development. Drawing and coloring activities also develop the child's perceptual knowledge. Beautifully colored, smooth and rounded things large enough to be grasped with a child's whole hand attract and maintain their attention.

Grade 2: This is a leveling off year.

Grades 3 and 4: Around the age of nine there is a noticeable change in the way the children approach reality. The line between reality and make-believe begins to be more firmly drawn. The boundaries of their emerging conceptions of reality are still being extended by active experimenting but there is a more real sense of what is possible or impossible, true or not true.

Grades 5: This is a leveling off year. More students are beginning to use adult reasoning in this grade in areas where they have had "real" experiences.

Grades 6,7 and 8: As children approach adolescence they begin to appreciate exactness in speech and in measurement. An interest in statements of rules or laws appears. Which are breakable and which are unbreakable? What is firm and hard?. Straight lines drawn with a ruler and smooth circles drawn with a compass suit their appreciation for more precision.

These descriptions brush broadly across the grades and are intended to help beginning teachers to focus on the individual children as they move through each stage of mathematical/logical development.

Developing Mathematical Concepts

Piaget distinguishes between "perceptual" and "representational" levels of knowledge in children. At the perceptual level, children derive their knowledge from direct contact with objects, whereas at the representational level, they are able to imagine new situations, draw inferences, and predict. Representational knowledge is, there, more abstract than perceptual knowledge since it moves the student to generalize to a whole class of situations. For example, a student tracing circular lids of various sizes notices that bigger circles have bigger diameters. This is perceptual knowledge. After measuring several different sized round lids, noting their circumferences and diameters, she notices that the ratio tends to be around 3 to 1 and she wonders if this ratio holds true for all circles. When she hypothesizes that the same relationship might hold for all round objects and experiments to check out her hypothesis, her knowledge gains "representational" status. She may represent this knowledge as a verbal or numerical statement e.g. "The circumference of a circle is about three times as long as its diameter" or $C = \pi d$.

In recommending different kinds of lessons in the primary and intermediate grades for the same essential understanding, the *Model Curriculum Guide* acknowledges these differing levels of knowing as they develop. Primary grade children are at the perceptual level in nearly everything whereas intermediate grade children are able to think at the representational level in some areas. Adults, also, are at the perceptual level of knowledge in many areas as well. Second-grade students who "discover" pi, as mentioned above, are not impressed by it the way sixth-grade students or adults are because they are not thinking at the same level. Giving them this kind of experience too early is like casting a seed on uncultivated ground: nothing much happens to it. Similarly, older students and adults who have missed out on the perceptual experiences involved in "playing" with circles may learn the formula for circumference but won't think of using it in situations where it applies. Perceptual knowledge is valuable.

Understanding at the representational level does not mean that a student can read or write mathematically. Learning to write an equation such as the one above for circumference requires that the teacher explain the social conventions that a group of people have developed over the years for writing these kinds of relationships e.g. writing pi next to d means that these numbers are multiplied. This kind of **social**

knowledge is different from the **logical/mathematical knowledge** involved in understanding the relationship between circumference, pi and diameter. Whereas social knowledge can change depending on the language and culture in which it is learned (different alphabet, mathematical symbolism, vocabulary, etc.) logical/mathematical knowledge remains the same no matter where it is learned.

Manipulatives

Manipulatives are materials that embody a mathematical concept in such a way that the learner mentally constructs concepts through manipulating the materials. The research literature consistently shows a pattern of effectiveness for using manipulatives to develop concepts. Parham (1983), for example, synthesized the results of scores of experimental studies of manipulatives and found them superior to less active approaches (such as boardwork) for improving achievement in virtually all math topics. In classes in which they are in regular use, these are commonly used manipulatives:

Unifix cubes	**Geoboards**
Base 10 blocks	**Color tiles**
Tangrams	**Grid paper**
Pattern Blocks	**String**
Attribute blocks	**Spinners**
Cuisenaire rods	**Counters (beans, macaroni,cubes, etc.)**

(Computers, calculators and number tiles are not considered manipulatives because they use symbols (e.g. "2" is a symbol) as the mode of communication with the learner.)

Although the research shows the effectiveness of manipulatives, their use is not yet widespread in schools. One impediment to the use of concrete materials in the classroom is their very attractiveness to children. Once children have access to concrete materials, their attention turns away from the teacher and toward the materials. Children can thus miss important information about the activity that the teacher has prepared for them and lose out on learning experience with manipulatives. A very helpful solution to this dilemma is to have materials distributed **after** the teacher is finished introducing the lesson. Designated students can do this and thereby free the teacher to supervise the whole group. We have also found that with a little practice children respond cooperatively in making the transition back to whole group instruction for summarizing if the teacher establishes a signal ahead of time. The next section compares exploring lessons with direct instruction and gives techniques for promoting active learning in both.

Preparing to Teach

Lesson Plans

As you observe several lessons being taught in the classroom, you will notice recognizable forms for each of these lessons. Some, although different in content and style of presentation, are in the same form: the parts of the lesson interrelate in the same way. Other lessons look different because they are in a different form. In this section you will see described two lesson forms which can be used to map out most of the lessons you may observe. The two lesson forms, **direct instruction*** and **exploring****, are the forms on which all the lessons in this book are based. The direct instruction lesson form is based on Madeline Hunter's lesson design format whereas the exploring lesson form was developed based on groupwork lessons emerging from the field of mathematics teaching. In addition, two variations of these forms - the **guided practice** lesson and the **student presentation** lesson - are also described in this section. The purpose of looking at "forms" is to provide a vocabulary for creating, discussing and analyzing rich and exciting math lessons.

Direct Instruction*	Exploring**

Anticipatory Set
Direct Instruction
 	
                    Check for
Input Model   understanding
 	
Guided Practice
Independent Practice
Closure
	Anticipatory Set
	Introducing
	   Explaining Activity
	
	   Organizing Groups
	       Roles of members
	       Rules for behavior
	   Establishing a signal
	
	Exploring
	Summarizing

There are philosophical and pedagogical differences between these two forms of teaching. The primary difference is that in direct instruction lessons, knowledge gains, which occur as students listen to and observe the teacher, are made firm by correct use during **independent practice** whereas in exploring lessons students do not practice what the teacher has taught them. Instead of practicing what the teacher has taught, students are put in a different situation in an exploring lesson. They are to learn from their own experience through **exploring,** with their peers, well-chosen problems and situations presented to them by the teacher. The next several pages explain and compare these two lesson forms further and give some practical suggestions for making them work.

Direct Instruction

Anticipatory Set
To focus student attention for instruction, a lesson needs a definite beginning. Asking a question, showing an object or explaining the usefulness of the lesson are examples of how teachers create anticipation which prepares student minds to learn.

Direct Instruction
Input/ Model/ Check for understanding
Instruction can be broken down by the microsecond using this format. What the teacher says (**input**), what she shows as she speaks (**model**) and how she determines whether the students are following her instruction (**check for understanding** or **CFU**) are noted in this way. For example, the teacher pictured at the right is finding out from her **CFU** (hands) that her instruction (**input**) is not reaching some of her students. She might make her  presentation more accessable, especially to visual learners, limited- English speakers, hearing-impaired, etc. if she would **model** some of what she says by using gestures, or using the board or overhead projector.

Guided Practice
In this part of a lesson, students begin work under the teacher's guidance. The teacher makes sure they understand which problems to do, how to figure them out, how to record the answers and so on.

Preparing to Teach

Independent Practice
Students continue the work independently as the teacher monitors and assists individuals as needed.

Closure
The teacher is again actively conducting the class. For example, answers are read or discussed, important points are reiterated, student work is collected.

Exploring

Anticipatory Set
Introducing
The introduction to an exploring lesson involves preparing students for the mathematical work they will do (the **activity**) as well as the complex social situation they will engage in (**groupwork**). To **introduce the activity:** present or review concepts involved, pose part of the problem or a similar but smaller problem for students to try and then present the problem to be solved. To **prepare for groupwork**: form the groups, assign each member a **role** and discuss the **rules** for behavior. Before beginning exploring, establish a **signal** (e.g. a bell) that indicates when exploring is over and it is time to be quiet and listen to the teacher. (You can analyze the way you do **introducing** using the **input/model/check for understanding** format in direct instruction.)

Exploring

Once students have settled into exploring, the teacher is free to observe the interaction and discussion of the groups. Information gathered at this time can be used during **summarizing** to guide class discussion. Offer assistance when needed and provide an extension activity for groups that finish before the others.

Preparing to Teach

Summarizing

Give the **signal** for attention e.g. the bell. Once the whole group is ready, begin the **summarizing**. Have groups report both on the **activity** and on the **groupwork**. Have them present their **solutions** and any **generalizations** they made.

Guided Practice Lesson

This is a common variation of the direct instruction lesson. In this form of lesson, students follow along with the teacher during the major part of the lesson; there is little or no independent practice.

Since the purpose of a guided practice lesson is for students to follow the teacher's line of reasoning rather than to generate their own, their job is to **listen** and **do as shown.** Conducting this kind of lesson, therefore, requires enthusiasm and stamina on the part of the teacher since she must keep the entire group with her as she explains.

The use of manipulatives during guided practice creates a tension in instruction: will students pay attention to the materials or to the teacher? Developing lessons in this form, therefore, requires developing, first, a very strong working relationship between teacher and students in order to work. Without this strong relationship, putting manipulatives in students' hands when they should be paying attention to the teacher may be working at cross purposes. A well-designed **exploring** lesson is often a less stressful vehicle for using manipulatives in instruction.

Preparing to Teach

Student Presentation Lesson

This can be a variation of **direct instruction**, as in the case when students are learning to use a standard mathematical form such as correctly writing a "missing addends" equation or a standard algorithm. Students practice using the conventional symbolism during the independent practice part of the lesson and share their written work and explanations with the whole class during **closure**. This lesson form can also be a variation of **exploring** when student groups, using a variety of approaches, strategies and problems, share their processes with the whole class during **summarizing**. Lessons such as *Word Problems in Multiplication and Division* and *Multiplying Decimals* in the Number chapter of this book are examples of student presentation lessons based on exploring.

In the lesson pictured below, student pairs have been solving a problem using either drawings, base-10 materials or "number crunching." During summarizing, representatives from each pair show the whole class how they solved the problem. Notice that this student presenter solved the problem with base-10 materials by starting at the "front end" of the problem first. The last step she does to get the sum of 162 is to exchange ten units for a long. Raising his hand is a boy who used a

different approach: number crunching. In an **exploring** lesson, the student's job is to **choose an approach,** solve the problem and then **judge** the soundness of each approach presented during **summarizing**. The teacher's job is to 1. provide a predictable format in which each student can present his work, field questions, and make way for the next student presentation 2. keep the environment safe for making mistakes, discovering mistakes, and correcting mistakes 3. in the case of exploring lessons, to encourage a diversity of approaches to the problem.

Speaking to 30+ Children

Input/ Model/ Check for Understanding

Communicating effectively to 30+ children for six hours a day is both a fine art and a prodigious practical science. A broad repertoire in content is essential, of course. The **input** or what you say to children shows your understanding of the content, the children, and importance you attach to them. The first time you teach a new subject, your delivery will be rough so you will need to rehearse it. Writing it out serves as a kind of rehearsal. Describing your plan of instruction to another person is a kind of rehearsal, too. The best kind of rehearsal, though, is to get to **teach it more than once** (e.g. to several small groups, to a friend or to half of the class and then the other half). This will allow you to make your input more effective by showing a **model** and by developing a repertoire of **check for understanding** techniques. How do teachers check for understanding?

The two most common techniques teachers use to **check for understanding** are: calling on **volunteers** - students raise their hands and the teacher calls on them one at a time, and **call outs** - students call out answers as they are ready. Using volunteers makes for an orderly and thoughtful recitation. Students listen and speak in their turn. It can become frustrating, though, when the whole class gets excited and everyone who has something to say is not called on to speak. Using call outs, on the other hand, makes for a more lively recitation and allows everyone to speak. It can easily get out of hand, however, so teachers use it with restraint. So then, different situations call for different CFUs. What are other means that teachers use, **as they speak**, to gauge the understanding, interest and motivation of the whole class? Here are several whole group techniques.

CHECK FOR UNDERSTANDING TECHNIQUES

Pair Share: Students briefly explain or demonstrate a concept or skill to each other in pairs.
Finger signs: Students answer by showing a number on their fingers.
Hand Signals: Students agree or disagree with an answer, e.g. by showing thumbs up or down.
Slates: Students write an answer on a slate and show it to the teacher.
Point at word/picture/number in book: Pointing shows if they are with you and ready.
Response cubes: Student show the numerals through 12 on these large, wooden dice.
Show number on manipulatives: Students hold up their manipulatives or leave them on their desks to show a number, a shape, etc.

By providing such means for students to be active, teachers can keep students involved and also get immediate feedback on their instruction. A very interesting study by Pratton and Hales (1986) shows the potential for a well developed repertoire of CFU techniques to keep students actively involved in the instruction. Twenty classes of fifth-graders were taught a unit on probability. Ten classes were taught using non-active participation (e.g. the teachers used boardwork to explain how to determine probabilities as the children watched) while the other ten classes used active participation (e.g. the teachers used several check for understanding techniques during instruction). The table below shows the results in achievement.

A COMPARISON OF ACTIVE AND NON-ACTIVE LEARNING IN PROBABILITY

Means and Standard Deviations of the Posttest Scores by Class

Students Active		Students Non-Active	
Class Mean	Class Standard Deviation	Class Mean	Class Standard Deviation
12.41	1.72	10.69	3.20
11.76	1.89	11.33	2.75
12.67	1.87	11.15	2.82
12.79	1.37	11.14	2.95
12.53	1.75	11.25	2.07
12.65	1.85	11.48	2.44
12.89	1.47	10.91	2.68
13.08	1.35	11.08	2.86
12.13	2.17	11.55	2.78
11.83	2.53	11.00	2.13

From Pratton, J. and Hales, L. (1986)

The active participation groups were more successful in achieving significantly better understanding with less variability in achievement. Keeping the whole class active during the teacher's recitation is one way to enhance learning. Another way is to keep them active by letting them work in groups.

Groups and Roles

Arranging children into groups to explore with manipulatives, write and discuss puts children into social situations requiring complex social skills. Left without a teacher's intervention, some children will complain about who is in their group. Once in the groups, some argue over who gets to do what. Some students will dominate in a group while others will withdraw and not participate. Teachers anticipating these problems can ameliorate them by thinking ahead of time about how to compose the groups and what jobs or roles students should have in the groups.

Our experiences in the primary grades indicates that children in grades one through three benefit from groupwork in a less complex social group than intermediate grade students. Working in pairs seems to work best for this age of children. Kindergarten-aged children seem generally to do better in a small table-group with a teacher or aide to supervise. In *Make It Simpler*, Meyer and Sallee give suggestions for organizing groups of four in middle and upper elementary grades to do problem solving.They recommend assigning children to groups by distributing a deck of cards and arranging the children with like numbers into groups of four. So a class of 32 would have eight groups whose membership would change every few weeks with another shuffling of the cards. Some of the student teachers who taught the lessons in this book were in classrooms in which there was no such longterm groupwork to allow for reshuffling. Since their time was limited they chose, instead, to carefully compose each group of four by consulting with their cooperating teacher. What did they consider in composing these groups? The ability mix (either heterogeneous or homogeneous); personality characteristics (domineering or shy, for instance); sex and race (balance in each group); and placing "problems" (children with behavior problems or learning problems). Whether or not you are in a classroom which uses groupwork will affect the smoothness with which your groupwork lesson will be carried out.

Once in groups, all students should take part in the activity at hand. Giving everyone a job gives legitimacy to the roles that each person takes. This is especially critical for students who are shy or quiet, limited- or non-English speaking, handicapped or in some way marginalized by the other students. Assigning roles has the effect of assigning status to students who might otherwise have little status on their own. Dividing up the work this way also distributes the weight of responsibility among the members. Meyer and Sallee, for instance, describe four roles that need to be filled for successful problem solving in groups. Here are their titles and typical comments.

Roles for Problem Solving*

Questioner "What are we supposed to do?" "Where are the scissors?"

Doer "I'll get the blocks, you get the markers." "Let's divide the task up into parts." "All right, everybody - let's get started."

Prober "I wonder if this will produce a pattern." "If we used triangles instead of squares, we might...." "I want to find out if this is true for all rectangles...."

Summarizer "Now let's review the directions." "Remember to put your name on your paper." "Oh, that's a nice design, John. Don't you think it is interesting, Maria?"

---

* From *Make It Simpler* by Meyer and Sallee

Other titles such as **Getter, Recorder, Writer, Encourager, Coordinator, Materials Manager** and so on can be invented for a given activity, and described and modeled for students as they prepare for groupwork.

It is *extremely* beneficial for students to reflect on what went on during groupwork as they are getting used to it. The sheet "Group Processing" in the appendix will help them to do this. Successfully solving problems of all kinds in groupwork will give long-term benefits not only in school but also in the family and, later, at work.

Rules and Signals

Anyone who works with large groups of children soon learns the necessity of establishing norms for student behavior for keeping peace in the classroom. For example, during a class recitation the norms may be: listen quietly, raise your hand if you wish to ask a question or make a comment to the teacher, and do not speak to the other students. During a lecture the norms may be even more restrictive: listen quietly and take notes. But research on problem solving shows that more active involvement is necessary in order for children to become successful problem solvers. Groupwork, when structured well, provides opportunities for all children to have a more active role - in speaking, manipulating, and writing - than does a large group recitation. So, what are the norms for student behavior during groupwork? And how does the teacher communicate these norms?

It is very important to recognize that the classroom behavior norms for groupwork are **different** from the norms for a recitation. In an **exploring** lesson the rules for behavior are so different from the rules for direct instruction that teachers who regularly use groupwork will post the rules for groupwork on the classroom wall. The three simple rules are as follows.

RULES FOR GROUPWORK

You are responsible for your own work and behavior.

You must be willing to help any group member who asks.

You may ask the teacher for help only when everyone in your group has the same question.

You are responsible for your own work and behavior.
Students already know this rule but it is especially important to have it posted during groupwork because groupwork provides many opportunities for this rule to be tested. If, for instance, two children argue and one swats the other claiming,"He made me!" this rule puts the responsibility where it rightfully belongs. Class discussions about how to deal with disagreements ("getting along") are very beneficial. For intermediate grade children there begins to be an added dimension:

differing perspectives on an issue. The outcomes of such discussions are an important part of the social learning that also takes place as a result of groupwork.

You must be willing to help any group member who asks.
The second rule goes **against** the norm of the class recitation which normally requires students to work independently without talking to the other students ("No talking."). By requiring students to help each other, the rules of groupwork may seem like "cheating" to children who have been trained to listen quietly and work independently in the classroom. Let them know that **everyone in the class** has something to teach or to contribute in some way. In groups, every person has the opportunity to both help and to receive help. By learning how to ask good questions and to really listen to the answers, we become better learners. And when we explain what we know to others, we become teachers. Giving help and receiving it are what we get to do when we work together in groups.

You may ask the teacher for help only when everyone in your group has the same question.
The third rule also goes **against** the norm for the large group recitation which normally requires students to raise their hands and wait only for the teacher to answer any question they have. This third rule for groupwork is a lifesaver for the teacher. It eliminates repetitious procedural questions (e.g. "Where are we supposed to put our papers?") which can be answered by children in the group and it turns the more substantive questions (e.g. "Does this look like the same pattern as the paper tearing problem?") to the children themselves as they begin to take on a more responsible role within the group.

Other rules may be given as needed according to trends the teacher sees among the students (e.g. no "put-downs," stay in your seats, use 12-inch voices, etc.) Using these rules, a classroom of 32 students arranged in groups of four becomes a classroom of 8 groups, each with a developing infrastructure for learning at several levels.

Even in classrooms where groupwork is used regularly, most classroom discourse is done under rules that put attention on the teacher.

RULES FOR RECITATION

Be polite. (You are responsible for your own work and behavior.)
No talking. (Listen attentively to the teacher and do not talk to other students.)
Raise your hand. (For any questions, raise your hand and wait only for the teacher to answer you.)

Using groupwork introduces more complexity into the social setting of the classroom since students will be operating under different social norms (**rules for groupwork** or **rules for recitation**) at different times. How does the teacher signal to students that the norms for behavior are about to change? Teachers do this by using signals.

Signals

Signals are explicitly taught social conventions that regulate classroom discourse. Like traffic signals on a busy street, they regulate the ebb and flow of activity in the classroom. Signals are so imbedded in the way teachers carry on classroom discourse that they are usually invisible to an untrained observer. Mixed or unclear signals lead inevitably to miscommunication and friction between teachers and students whereas clear signals contribute to warm and lively communication every day in school.

Getting the group's attention and holding it is a very important part of what teachers want to be able to do and teachers have signals to indicate that norms are being broken during instruction:

SITUATION	SIGNAL	CHECK FOR UNDERSTANDING
Some children are talking while teacher is talking to the class.	A long pause, a silent stare or a gaze in the direction of the talking.	Students notice the silence and become silent.
	Proximity: Teacher stands next to the talkers.	The talkers notice that the teacher has noticed them and become quiet.
	Announce name(s) of talker(s) e.g. "Kelsey."	Talking stops.
	"I'm waiting."	They quiet down.
	"It's my turn."	Students wait for "their turn."

Teachers also use signals to indicate when to begin groupwork or independent practice and when to end it. The following are two lists of some common verbal and nonverbal signals used by teachers in active classrooms to signal a change in norms. The first set of signals, **Releasing Them**, releases students from the teacher's direction to begin either independent practice or exploring work. The second set of signals, **Getting Them Back**, signals students to return to the teacher for directions. Once students understand the meanings of these signals, they can help teachers to maintain positive management in the learning environment.

RELEASING THEM

SITUATION	SIGNAL	CHECK FOR UNDERSTANDING
Children have been listening to teacher's instructions and it is time to start work.	"The 'red light' is on."	There is no talking as the class begins to work independently.
	"Begin."	Quiet work begins e.g. independent practice, "seatwork," a test.
	"Ready, go."	A timed or speeded activity begins e.g. a race, game, timed test.
	"Begin groupwork."	Students turn from the teacher to each other and begin to talk.
	"Take one minute to 'pair-share' with your neighbor how you would __ (solve this problem, explain this process, feel in this situation)."	Students talk quickly in pairs for one minute.

Preparing to Teach

GETTING THEM BACK

SITUATION	SIGNAL	CHECK FOR UNDERSTANDING
Children are discussing in groups.	A bell rings. (Or a timer, piano chord, xylaphone, etc.)	They stop and listen.
	"All eyes on me."	Children look at you.
	Lights off.	Work stops.
Children are talking informally.	**Teacher stands in a particular location** in the classroom.	**Students come to order for instruction.**
	Hand up two fingers in the air.	Hand up and quiet two fingers in the air and quiet.
	Call and answer. e.g."When the hand goes up...."	The mouth goes shut!"
	"Brain alert!"	Quiet and ready for something requiring thought.
	Echo clapping: e.g. Clap-clap, clap-clap-clap.	Clap-clap, clap-clap-clap. (Echo)
	e.g. Clap-clap, clap-CLAP-clap...	CLAP-CLAP. (Finishes it.)
	"Class!"	Talking stops. Class attentive.
Children are using manipulatives.	**"Hands in your lap."**	**Hands off manipulatives and listening.**
	"You have one minute to finish what you're doing."	Increased activity as they prepare to stop activity.
	"Put your (cubes, beans, base-10 materials, etc.) back into their bag and look at me when you're ready."	Students begin to clear manipulatives from the desk and put them into a bag.
Children working independently or in groups.	**"You have one minute to finish what you're doing."**	**A flurry of activity to complete individual or group work and** prepare to share it.
	"Group leaders, make sure your group is ready to share its findings."	Group leaders make final arrangements for summarizing findings.
Children are out of their regular assigned seats.	" 10, 9, 8,...2, 1, 0"	**Return quickly to their regular assigned seats by end of count.**
	"Take your seats."	Return to regular assigned seats.

Geometry

Essential Understandings from the Model Curriculum

It is essential for students to understand that:

1. Real objects and abstract figures have one, two, and/or three dimensional features which can be examined, compared, and analyzed.

2. Geometric figures have specified attributes and properties by which they are identified, classified, and named.

3. Geometric figures can be described in terms of their relationships with other figures. Important relationships include relative size, position, orientation, congruence, and similarity.

4. Geometric figures can be composed of or broken down into other geometric figures.

5. Relationships within and between geometric figures can be revealed through measuring and looking for patterns. Constant relationships can be expressed as formulas.

Preparing to Teach

Developmental studies of children's geometric concepts suggest a particular order of development: topological concepts, then projective concepts, and finally, Euclidean concepts appear. A good understanding of Euclidean concepts depends on the orderly development of the two previous groups of concepts. You will see children's interests from kindergarten through grade eight progress in this order. Here is a description of what you will see.

In **topology**, young children are interested in how shapes can be altered by bending, stretching, twisting, and compressing but not by tearing or joining. Some properties of shapes that stay the same while they are being changed in these ways include open, closed, simple, or non-simple. An activity in which children can explore topological ideas is to have them sketch simple figures on a balloon and observe the figures as they stretch and pull the balloon.

Projective geometry is shadow geometry and older children can explore this inside or outside of the classroom. Of the two kinds of projective geometry, the first is what results from using a point as the source of light, as in the case of shadows cast using a flashlight and the second, called *affine geometry*, results when rays of light are approximately parallel as in the case of sunlight. The shadows that can be obtained are different in some respects using these two sources. In affine geometry, for example, a shape with parallel sides will always cast a shadow with parallel sides, whereas in point-source projections this will not necessarily happen. This difference can be verified by comparing how shadows made outside in the sun compare with shadows inside the classroom from a flashlight. A particular case of projective geometry that should be mentioned occurs when a shape is held parallel to the surface on which its shadow is being cast. In this position, the shadow will be similar to the shape. An activity for children involving point-source projections is to compare a slide photography with its image on a screen. An good affine activity is to use a shadow technique to find the height of an inaccessibly tall object outdoors. Children learning about proportion are impressed to find that by measuring an object and its shadow, they now have the ratio for all objects and their shadows at that time. This information can be used to calculate at heights without measuring them.

Older children can also appreciate exploring **euclidean geometry** which deals with the properties of shapes that stay the same as the shape is moved around through a reflection, rotation, or translation. As these transformations alter neither the size nor the shape of figures, such properties as lengths of sides and measures of angles remain unchanged. The ideas of congruence are studied in euclidean geometry. Most teachers are very familiar with activities that use reflections, rotations, or translations. Quite a lot of the teaching of geometry draws to some extent on these ways of moving shapes. For example, lessons on line symmetry are common, and reflection is often taught through paper-folding or mirror techniques.

Children whose concrete experiences with geometric shapes, such as triangles, have been limited develop misconceptions. For instance, when some children were asked to draw five different triangles, a common response was to draw an equilateral triangle and then to turn it in various directions to make it "different." Or, they drew one side with wavy lines to change it. In another example, children were asked to circle shapes that looked like triangles. Shapes that looked like

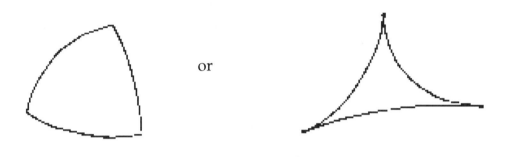

were frequently circled but shapes that looked like

were not. These examples illustrate the importance of concrete experience in developing concepts.

Among children who are aware of the properties of the components of shapes, many still do not use the properties to define the shapes. For example, to many children squares are not rectangles and rectangles are not parallelograms, even though they share the same properties. Such beliefs persist among many high school students, even those who memorize correct definitions of the shapes. These findings point out the need for a realistic curriculum in geometry in the grades preceding high school.

How do students at different levels of experience reason differently in geometry? Dina van Hiele (Burger and Shaughnessy, 1986) found these five discrete levels of development in Euclidean geometry.

Preparing to Teach

> **The van Hiele Levels of Development in Geometry**
>
> **Level 0 (Visualization).** The student reasons about basic geometric concepts, such as simple shapes, primarily by means of visual considerations of the concept as a whole without explicit regard to properties of its components.
>
> **Level 1 (Analysis).** The student logically orders the properties of concepts, forms abstract definitions, and can distinguish between necessity and sufficiency of a set of properties in determining a concept.
>
> **Level 2 (Abstraction).** The student reasons about geometric concepts by means of an informal analysis of component parts and attributes. Necessary properties of the concept are established.
>
> **Level 3 (Deduction).** The student reasons formally within the context of a mathematical system, complete with undefined terms, axioms, and underlying logical systems, definitions, and theorems.
>
> **Level 4 (rigor).** The student can compare systems based on different axioms and can study various geometries in the absence of concrete models.

Here are examples of the kind of responses that students of the different levels would make to the questions "What type of figure is this? How do you know?"

What type of figure is this?

Students at each level are able to respond "Rectangle" to the first question. (If a student does not know how to name the figure, he or she is not at level 0 for rectangles.)

Students' answers to the second question reveal the kind of reasoning they are using to answer the question.

How do you know?

Children in kindergarten through grade six primarily span from level pre- zero to level two or (in some cases) three. Here are examples of the kind of responses that students of the different levels would make to the second question.

Level 0: "It looks like one."

or

"Because it looks like a door."

The answer is based on a visual model.

Level 1: "Four sides, closed, two long sides, two shorter sides, opposite sides parallel, four right angles. . ."

Properties of the rectangle are listed but the redundancies are not seen.

Level 2: "It is a parallelogram with right angles."

The student attempts to give a minimum number of properties. (If queried, she would indicate that she knows it is redundant in this example to say that opposite sides are congruent.)

Level 3: "This can be proved if I know this figure is a parallelogram and that one angle is a right angle."

The student seeks to prove the fact deductively.

Research has supported the validity of this model for assessing student understanding of geometry and for guiding the design of materials and methods that promote growth through levels of understanding.

What might a sound program in geometry look like in kindergarten through grade eight? The program for geometry at the Sacramento Waldorf School is part of an entire school curriculum designed in the early part of this century by the German educator, Rudolph Steiner. The curriculum in geometry offers many desirable features based on what we now know about development of understanding in geometry. The next few paragraphs describe briefly the emphases in the various grades.

Geometry at the Waldorf School

In the primary grades at Waldorf Steiner also believed that geometric shapes should be experienced in a physical way by the child.

". . . the teacher should not be satisfied when the children can draw a circle - our children must learn to feel the circle, the triangle, the square. They must draw the circle in such a way as to have the roundness in their feeling. They must so learn to draw the triangle so that they have the three corners in their feeling, that already in drawing the first corner they have the feeling that three corners are emerging. In the same way they draw the square so that they feel the emergence of angularity, that the feeling permeates the whole line-development right from the start. A child has to learn from us what a curve is, what a horizontal line is, what a vertical line is, but not merely from his observation, but from following it inwardly with his arm, with his hand."

Artistic form making of the shapes characterizes the geometry program in the primary grades for children up to the age of nine.

From age nine up to the ages of eleven or twelve, the program can be characterized as "descriptive observation." Beginning in the third or fourth grade children begin drawing both two- and three-dimensional shapes with more exact lines. Projective geometry and perspective drawing form the fifth-and-sixth grade geometry units.

Sometime during grades six, seven and eight a study of the angle sum theorem is made. This is the first serious entry into Euclidean geometry and it is the goal toward which the earlier experiences with shapes are aimed. Steiner believed that with the proper background experiences children could genuinely understand the theorem in a short time:

"I have often said, provided we choose the right age we can teach young people in three or four hours from the beginning of geometry, the straight line and the angle, to the [angle sum theorem] . . .But think for a moment what rubbish goes on in present-day teaching before people reach this theorem! The point is that we have wasted a tremendous amount of. . .effort.. ." (Stockmeyer p. 78)

Though great progress has been made in instruction in geometry since 1919, for many children what Steiner said is still true. Burger and Shaughnessy (1986) found many students who had completed study of geometry in high school who were assessed formally at level 0 or 1 on tasks, not level 2 or 3 as might have been expected.

These geometry lessons suggest a progression from kindergarten through grade eight. Look them over and try teaching them. What levels of understanding do you find in the various grades? What does the geometry program in you school look like?

Measurement

Essential Understandings from the Model Curriculum Guide

It is essential for students to understand that:

1. When we measure, we attach a number to a quantity using a unit which is chosen according to the properties of the quantity to be measured.

2. Choosing an appropriate measuring tool requires considering the size of what is to be measured and the use of the measure.

3. Measurement is approximate because of the limitations of the ability to read a measuring instrument and the precision of the measuring instrument. The more accuracy you need, the smaller the unit you need.

4. For an accurate drawing or model to be made, a constant ratio between the lengths of the model and the lengths of the real object must be maintained.

Preparing to Teach

Much of the instruction in measurement that currently takes place in schools deals directly with computation in using and converting standard measurements. For example, we teach children how to convert feet to inches and how to add and subtract hours and minutes. Concept development is given little attention generally because we assume, correctly in some cases, that measurement concepts develop outside of school. Children learn at the grocery store, for instance, how big a liter is and how heavy five pounds of potatoes are. What aspects of size or quantity should children in grades K through 6 learn? There are five:

> volume
> weight
> length
> area
> temperature

All of these are qualities of physical objects. An awareness of these qualities arises as the child handles various physical objects. Once a child perceives these qualities, how does this become conceptual knowledge? It develops in this way:

- Comparing size or amount (e.g.bigger, smaller)
- Measuring using non-standard units (e.g. using a hand)
- Measuring using standard units (e.g.using a foot)
- Choosing appropriate units for a particular measurement (e.g. deciding whether to measure the area of a wall in yards or square yards.)

Of these five qualities of objects, one concept with very rich potential for future learning in mathematics, if developed well, is the concept developed in the measurement of **area**.

Questions to ask ourselves as we teach in this strand are:
Where are children learning these concepts? What experiences can be provided through the home and which at school?

Probability and Statistics

Essential Understandings from the Model Curriculum Guide

It is essential for students to understand that:

1. When there is no direct observation that will answer a particular question, it is often possible to gather data which can be used to answer the question. Working with the data often generates additional questions.

2. You can gather data about every member of a group, or you can use a representative sample from that group.

3. Data can be organized, represented, and summarized in a variety of ways.

4. There are many reasons why an inference made from a set of data can be invalid.

5. There are ways to find out why some outcomes are more likely than other.

Preparing to Teach

The work of Piaget and Inhelder on the development of probabilistic thinking in children indicates three stages of development:

Stage 1 Children up to the ages of 7 or 8 do not understand random phenomena but continually look for a hidden controller. Their ability to make predictions is unstable, depending on personal preference, or on an expectation that one outcome will "catch up" with another, or simply based on the most frequent outcome observed. The occurrence of extremely unlikely events ("miracles") causes no surprise.

Stage 2 Given the appropriate experiences in the primary grades, children around the age of 9 (usually about third or fourth grade) change in the way they are able to think about probabilistic events. A global understanding of randomness is achieved. "Miracles" cause a search for an underlying cause. At this stage, however, the effect of large numbers in not yet understood. For example, children at this stage still may not understand that flipping a coin 30 times will give a better approximation of the true probability of getting heads than will flipping a coin only 10 times.

Stage 3 Given the appropriate previous experiences with random and non-random events, many children around the age of 13 will begin to think formally about probability. A rudimentary understanding of the law of large numbers is achieved and the probabilities of simple events, such as flipping heads on a coin, may be successfully assigned.

Children in the intermediate grades have an intense interest in notions of fairness and in games, particularly in how to win. The study of probability develops concepts and methods for investigating such situations. Methods of sampling allow students to make predictions when uncertainty exists and to make sense of advertised claims that they see and hear in the media. Investigations into probability also provide for the development and application of concepts of ratios, fractions, percents and decimals in middle or junior high school.

Many of the lessons in this chapter deal with statistics - collecting, graphing and interpreting data. Collecting and examining information on themselves is of great interest to even the youngest children. Real graphs using real objects such as shoes or cookies are made in kindergarten and first grade. Picture graphs using cutouts or drawings of the real things such as eye colors or ice cream cone flavors are excellent in all primary grades. Symbolic graphs such as bar graphs or line graphs are appropriately used in the upper elementary grades.

Our experience with the probability lessons in this book shows that most adults stand to gain a much deeper understanding of probability through actually doing these activities themselves. Even experiments shown for primary grades can powerfully convey concepts beyond what verbal descriptions can.

these activities themselves. Even experiments shown for primary grades can powerfully convey concepts beyond what verbal descriptions can.

Research on children's learning of probability and statistics supports the use of experiments but indicates that teachers may find that there is not much transfer of learning from task to task. For example, children who learn the concept of a random event using a die will not automatically transfer the concept when confronted with a random event using spinners. A good math program in this strand will provide a variety of instructional activities using spinners, dice, colored cubes hidden in a bag, beans in a jar and so on in developing probability concepts appropriate to the developmental level of the children.

A recommended approach to beginning the experiments in probability is to have the children predict what the outcome will be before having them begin. This arouses their interest and focuses their attention on the experiments. Several of the student teachers reported that in some classes children would erase erroneous predictions once the actual results were in so as to make the predicted and actual outcomes the same. If you are in a class that seems to expect "right answers" exclusively in math you may see this kind of behavior as well. Encourage the children to be like explorers or scientists. Making wrong guesses is allowed. After they have done an experiment several times their predictions will get better. Point out to them that testing our predictions is a valuable way to learn.

Patterns and Functions

Essential Understandings from the Model Curriculum Guide

It is essential for students to understand that:

1. Identifying a rule that could have been used to generate a pattern enables one to extend that pattern indefinitely.

2. When there is a functional relationship between two quantities, the value of the first quantity determines the corresponding value of the second.

"Looking for patterns helps bring order, cohesion, and predictability to seemingly unorganized situations. The recognition of patterns and functional relationships is a powerful problem-solving tool that enables one to simplify otherwise unmanageable tasks and to make generalizations beyond the information directly available."
(Model Curriculum Guide in Mathematics, 1987)

By looking for and finding patterns in math, children gain two of their most valuable treasures: a genuine, well-founded belief in the orderliness of mathematics and a confidence that through their own exploration they can expect to discover order and beauty beyond what they already know. The lessons in this chapter are designed to give students tools to do this. The understandings of the *Model Curriculum Guide* are developed in this way. Children notice existing patterns or create their own. They copy or recreate a pattern in a different form or mode through clapping, coloring, chanting, graphing or making a table. Older children can begin to write simple functional equations that a succinctly capture pattern with numbers.

A goal of patterns and functions is for children in the middle or junior high school to summarize data in these three ways:

>table
>graph
>equation

These lessons are rich in mathematical nuggets. Enjoy prospecting!

Logic

Essential Understandings from the Model Curriculum Guide

It is essential for students to understand that:

1. Classifying and sorting depends on the identification of a specific attribute or attributes.

2. Statements made precisely about what is known allow conjectures and conclusions to be examined logically.

3. Based on certain premises, a series of logical arguments can be used to reach a valid conclusion.

More than any other period of life, the seven years spanned by grades K-6 show the most rapid and noticeable changes in a person's logical thinking ability. Piaget describes a change from pre-operational to concrete operational to formal operational thought in this short period of time. Although empirical research with children of this age has shown that their progress through these stages is not unilateral across the strands but is dependent on the particular experiences the child has within each strand, developmental changes are clearly evident to the teacher who works in the various grades. It is this rapid change in the quality of the child's thinking that makes designing an integrated and appropriate mathematics curriculum for kindergarten through grade six so challenging. The *Arithmetic Teacher* does not even index the topic of logic in its annual index, it is so much a part of the math curriculum.

A major difference between the pre-operational child and the concrete operational child is that the concrete operational child can **conserve** which means he understands that when objects are changed in some ways certain properties of the objects remain the same (or are conserved) even though they appear different. For example, if you count a row of five pennies in front of a concrete operational child and then spread out the pennies, she will know that the number of pennies stayed the same even though there appeared to be more when they were spread apart. The pre-operational child has yet to realize what qualities are invariant when objects are transformed to appear different. Children achieve the ability to conserve in some part of the math curriculum by the age of seven or eight.

A major difference between the concrete operational child and the formal operational child is that the formal operational child is able to think hypothetically about a real situation and come to a rational conclusion about its outcome without using a physical model to understand the situation. For instance, a formal operational child would be able to identify what a shape would look like if it were rotated ninety degrees whereas a concrete operational child would need to physically rotate the shape to see what it would look like. Most adults are at the concrete operational level in most of the strands in math.

The *Model Curriculum Guide* describes appropriate activities for primary and intermediate level children in each of the strands. These are the guides for instruction in the logic strand:

Preparing to Teach

1. Classifying and sorting depends on the identification of a specific attribute or attributes.

<u>K-3</u>   <u>3-6</u>

a. Objects are physically present.
b. Any particular group of objects sorted in more than of attributes. one way.

a. Use more than one attribute.
b Use specific   definitions can be

2. Statements made precisely about what is known allow conjectures and conclusions to be examined logically.

K-3   3-6

a. Begin to use accurately ALL, SOME, NONE, EVERY, OR, AND, MANY.

a. Continue to use this vocabulary and add the logical form IF... THEN.

3. Based on certain premises, a series of logical arguments can be used to reach a valid conclusion.

K-3   3-6

a. Test their expectations against reality. false.

a. Consider how to show whether something is definitely

b. Can expect certain results all the time but in other situations the results are unpredictable.

b. Begin to verbalize steps in their reasoning.

Number

Essential Understandings from the Model Curriculum Guide

1. Numbers can be used to describe quantities and relationships between quantities.

2. Any number can be described in terms of how many of each group there are in a series of groups. Each group in the series is a fixed multiple (the base of the place value system) of the next smaller group.

3. The operations of addition, subtraction, multiplication, and division are related to one another and are used to obtain numerical information.

4. The degree of precision needed in calculating a number depends on how the result will be used.

The number strand is the part of the mathematics curriculum also referred to as arithmetic. It includes these topics: beginning number concepts, place value, basic facts, algorithms, fractions, decimals and percents.

Preparing to Teach

Beginning number concepts:
Children develop the beginning number concepts for the numbers 0 through 9 by sorting things into groups of different sizes, by holding things in their hands and noticing that some groups have more and some have less, and by counting and talking about how much they have. Once numbers are part of their speaking vocabulary, young children are ready to learn to read and write the symbols that stand for the numbers. As adults we can appreciate how the written words of a foreign language seem dead to us, appearing as unusual shapes with no meaning. For young children without concrete experiences, it is not just that the symbols seem dead. For them, the symbols are dead since there are no ideas to attach the symbols to, no "mother tongue" that they can translate to. The natural talk arising from concrete experiences with numbers makes the mother tongue of math.

Young children must count real things in order to develop beginning number concepts. And they will count real things even when nothing is provided for them to count. They will count their fingers and, if that is forbidden, they will nod their heads and count the nods. Kindergarten and first-grade teachers, then, have the formidable task of providing concrete experiences for children with these essential numbers.

Place value
The base-10 number system that we use provides us with an extremely efficient way of writing numbers of any magnitude. When we begin to understand place value, we begin to tap in to a way of thinking about numbers that has tremendous mathematical power. Both children and adults stand to collect great dividends from investing time in learning about place value. Using base-10 materials such as money ($100, $10 and $1), FLU materials (flats, longs and units), or bean sticks (made with craft sticks and beans of any kind) children lay the conceptual groundwork for learning the algorithms.

Basic facts
The basic facts for addition, subtraction, multiplication and division are what every elementary school student is expected to know by the end of the eighth grade. Recall of these facts will eventually be automatic provided that students receive these three things:

1. Information on how to reduce the number of facts to remember
With 81 facts to learn for each of the four basic processes it would seem impossible to expect anyone to learn all of the basic facts. But by using manipulative to show the facts, students can be shown the properties of the numbers that drastically reduce the number of different facts they have to learn. The commutative property, for example, immediately reduces the number of facts by almost one half.

2. Regular use and practice
Board games, class games, computer activities are some ways to give the needed practice.

3. Knowledge of results or "feedback" on their answers
This can be provided by the teacher or other students.

Algorithms
The algorithms are the step-by-step processes by which we do addition, subtraction, multiplication and division with two-digit numbers and beyond.
The "standard" algorithms taught in the schools, for example the one for long division, differ somewhat from country to country. Yet universally, many children experience difficulty in learning to use them correctly in school. A review of studies from South America (Nunes et al,1993) comparing the mathematics success of tradesmen using their own mental calculations as compared to school-taught routines for addition and multiplication, for example, showed interesting results. They did better using their own mental calculations. It appears that there is untapped potential in self-constructed mental mathematics which traditional instruction thwarts. The lessons you will see in this section should allow you to observe some of the thought processes of children as they grapple with computing with bigger numbers.

Fractions, Decimals, rates, ratios and percents
Understanding rational numbers is the big step that upper elementary students take into adult thinking. Students must learn to name, compare and calculate the algorithms. Concrete experiences are especially important since the numerals for the whole numbers mean something different in decimals and fractions. For example, 6 is less than 35 but .6 is more than .35. 16 is bigger than 2 but 1/16 is smaller than 1/2. A variety of experiences is provided in this final section.

CHAPTER 2
Geometry

Things that Stick Out in the World   Identifying solid shapes K-4
Bubble Magic   Surface area K-6
A Lesson Design in Geometry   Identifying solid shapes   K-3
Shapes with Felt and Yarn   Shape recognition K-3
The Surprising Circle   Circle quartering 3-8
Alphabet Symmetry   Finding lines of symmetry 4-6
Open and Closed Paths   Identifying open and closed spaces 2-4
Sorting Shapes   Classifying shapes 4-6
Triangle Designs   Developing angle concepts 4-8
Similarity and Congruence   Identifying similar and congruent shapes 5-8
Rotational Symmetry   Making symmetric designs 4-8
Polyhedrons   Constructing polyhedrons 4-8
Battleship Game   Graphing 4-8
Circles   Discovering pi 5-8
Projective Geometry   Changing the shape and size of shadows 6-8
Triangle Inequality and Triangle Nests   Properties of triangles 6-8
Fun and Folklore with Tangrams   Making tangram pieces 5-8
Tesselations   Making tesselating patterns 5-8
Geometric Quilt Bulletin Board   Making a geometric quilt K-8

Things that Stick out in the World

by Barbara F. Castrillo

An Introduction to Solid Geometry

Identify solid geometric shapes

Materials

- Solid geometric shapes to correspond with the shapes on worksheet
- Six covered shoe boxes with small flap cut in one end for "feeling"
- Worksheet (attached)

Introduction

Today you are going to discover what is inside these covered boxes on the table without looking inside. We have been doing a lot of activities with geometric shapes (squares, triangles, circles and rectangles) and now you are ready to use what you already know to discover what solid geometric shapes really feel like. Before we begin our exploration of the mystery boxes we need to review some of the vocabulary we have learned and several new words that you will need for your discovery.

Procedure

Have the materials member of each group pass out the attached worksheet, the same sheet is displayed on the overhead projector. Next, the teacher pulls similar objects (juice can, tennis ball, oatmeal box, cereal box, etc.) out of a bag behind the projector and describes how it "feels", the number of faces, what kind of edges and what shape the various surfaces have. At this point do not mention the labels listed on the worksheet.

Exploration logistics

Each of the numbered mystery boxes contains one of the solid geometric shapes pictured on the worksheet. One box will begin in each group. After each member of the group has felt inside, he/she will mark the number of the box next to the picture of the object he/she thinks is inside. After every member of the group has "voted", switch boxes with the adjacent group. For example, groups 1 and 2 switch boxes with each other, three and four, and five and six. This exchange of boxes will continue in a clockwise direction until each group member has felt and voted on each of the boxes. When the teacher gives the signal, each group will have one minute to finish and return the boxes to the front table.

Extension activity

After your group has finished "voting", turn your papers over, close your eyes and visualize the shapes you just felt. Then open your eyes and draw a picture of the shape that felt the best to you in the space provided on the worksheet.

* It is very important that everyone understands the procedure before the exploration begins, so a quick review of the box exchanging, recording and extension instructions is a good idea.

Let the exploration begin

Proceed as above. However, the teacher should give the signal (e.g. You have one minute to finish exploring) in plenty of time to allow for the ten minute "unveiling" during the summary.

Summary

Without showing the box number, the teacher reaches in one at a time and describes what he/she feels inside. Before opening each box have students show with their fingers, which box number they think you are holding and tally the responses. Teacher opens box, displays object and then gives the label for it that is listed under the picture.

If there is time let students share which shape they liked best and why (graph results). Ask students to think about how the picture of the shapes on the worksheet is different from the actual shape they felt inside the box. Challenge students to look for objects at home that have similar shapes and encourage them to bring some in to the classroom.

Follow-up activity

Let students know that the next day they will be creating special kinds of three dimensional objects at school and for this activity they will have to wear shorts or clothing they will not mind getting wet! **See next lesson on . . . . . BUBBLE MAGIC.

Teacher role

The teacher's primary role is to provide clear instructions before exploration begins to ensure the proper atmosphere for discovery learning. Then during the exploration, the teacher should move around the room, observe and offer assistance when needed (only if requested by all members of the group or when an obvious problem exists).

Assessment

Hidden Assessment: Children work in pairs as teacher surreptitiously assesses. One child describes the shape she feels inside the box as the second child tries to guess it. Teacher is assessing the first child. Then they switch roles and teacher assesses the other child.

Geometry

Name: _____

Shape	Box Number
sphere	
cube	
pyramid	
cone	
cylinder	
rectangular prism	

Write the box number for each shape.

My Favorite Shape

Bubble Magic

by Elizabeth Castillo

Students will have a wonderful time while learning about planes, surface area, and surface tension.

Materials

- Large Bucket
- Bubble Solution (2 cups Dawn and 3 T Glycerin)
- Bubble Wands (all shapes and sizes)
- Juice cans
- Kiddie Pool
- Milk Carton from Cafeteria (washed and dried)
- Hula Hoops
- Coffee cans
- Pipes
- Flexible Straws
- Strings

Procedure

Mix Bubble Solution 5 days before using for best results. The more helping hands available the better this lesson works and parents enjoy it as much as the children do! Each group of four should have one pie plate full of bubble solution and at least five assorted bubble blowing items (some listed above). This is an outdoor activity. However the teacher can model the procedure inside the classroom by demonstrating bubble blowing techniques. Remind the students that the bubbles are easier to work with if the tools (including hands) are wet with solution first.

Introduction

Teacher introduces the concepts plane (curved and flat) and surface tension. Demonstrate first by filling a glass of water to the "top" (a flat plane) and then use an eye dropper to keep adding drops until a curved plane is visible.* Then blow several bubbles. Four straws joined together with string form a cube shaped bubble blower, challenge the children to use the flexible straws and other materials available to make as many different geometric shaped bubbles as possible. Have six stations set up outside before introduction because each group will be anxious to begin the exploration time. Warn students that the ringing of a bell, or other signal, means they have five minutes to finish and clean up.

*The water doesn't spill out even though it isn't flat because of surface tension - a strong "skin" formed by the water.

Exploration

Teacher role is to walk from group to group to monitor and observe.
Please remember to take a camera to school (with film) the day you do this activity. The fewer constraints placed on this activity the better the bubble solution acts as a magnet for creative exploration. At the end of the activity, pour the leftover solution in the kiddie pool. Ask for a volunteer from each group to take off their shoes and stand in the pool. Pass the hula hoop from the pool up over their head, and they will be "enclosed" in a transparent cylinder. Repeat as time allows.

Summary

Let each group report on the results of their exploration, record the observations and comments on the overhead projector. Encourage the use of rich, descriptive language. Save the left over solution in a covered container and refrigerate.

Follow up

Creative writing assignment on what it would be like to travel inside a bubble. Encourage the students to become transformed. Have this be an "in class" assignment while the experience of creating the bubbles is still fresh in their mind and senses.

When the pictures come back that you took of the exploration activity, post them on a bulletin board along with some of the creative writing.

Assessment

Journal writing: What shape was your "bubble wand?" What shape bubbles did you get from it? Describe them and draw pictures.

Source: The March 1985 issue of *National Geographic World*, pages 120-121, "Bubbliology". For more specific procedural question, please refer to this article.

A Lesson Design in Geometry

by Linda C. Bolin

Identify shapes

Children will create their own designs using wood and paper shapes (e.g., hexagon, triangle, square, rectangle, and circle) and will describe their designs using the names of the shapes.

Materials

Geometric wood shapes of various colors, shapes construction paper shapes (same color and size as wood shapes), 4 1/2 X 6 black construction sheet, (blackboards), paste, and the geometric shapes enlarged and cut from large construction sheets the same color as the wood shapes
(see samples below).

Anticipatory set

Display large construction-made geometric shapes and say, "Today we are going to learn the name of these geometric shapes and will also create our own geometric designs."

INPUT	MODEL	CHECK FOR UNDERSTANDING
Boys and girls, each one of these shapes has its own name. Let's say each one. First my turn. (Say name and have group repeat.) Your turn. (Continue until all names have been said. Call on individuals until all have responded.) Good listening and saying geometric names.	Show display shape. 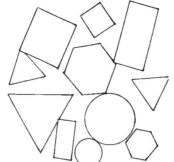	Group response

Geometry

INPUT	MODEL	CHECK FOR UNDERSTANDING
First you will use these wood shapes and see what different designs you can make and how you can change the shapes around to get a different design. Make your designs on your own blackboards; you may use any shapes you choose but remember, we are using only shapes because that is the number we are learning about this week. Be sure to keep your shapes flat on the paper because after you have used them and have made a design that you like, I want you to then make the same design on the black sheet using the paper shapes and paste them on. You may then take your design home and tell your family the names of the shapes and really impress them!	Display wood shapes. Show board and flat vs. upright. Display sheet and paper shapes.	
Let's go over the directions (Repeat and call on individuals to CFU) Good listening. When I call your name, you may sit at the table where a blackboard is and choose your shapes. (Call names.)		Individual responses. Each child sits at the table and begins practice.
When you have finished, please clean up your area and put your hands on your head and then I will know you have finished your design. (Students who have cleaned up may continue to use wood shapes.)		Hands on heads.
Everyone is finished. Let's all hold up our designs so we all can see them. Great designs!		Hold up sheets.

Practice

Each child makes a design using five wooden shapes and then copies the design on black paper using paper shapes and paste.

Summary

(Each student is to show finished design to teacher.) What shape did you use in your design? Tell us about your design.

Today we learned about geometric shapes, their names and how we can use them to make designs.

Assessment

Hidden Assessment: Have two children at a time work together as you surreptitiously assess them. Show them five designs from the classwork. The one who is "it" chooses a design from the five and gives hints to help her partner guess which design it. The hints include the names of the shapes and their relative position in the design. Then the pair switches roles. You assess the one who is "it."

Source: Mary Baratta - Lorton, *Math Their Way* (Menlo Park, California: Addison Wesley Publishing Company, 1976)

Geometry

Shapes with Felt and Yarn

by Elizabeth Castillo

Shape recognition

- The students will arrange their individual length of yarn into the shapes of a circle, square, triangle and rectangle

Materials

- Felt board
- Felt pieces of the colors red, green blue and yellow in the shapes of circles, squares, triangles and rectangles. Each shape has three sizes: small, medium and large.
- A 12-inch loop of yarn (closed with a piece of tape) for each child

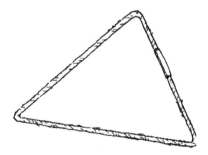

Introduction

(The teacher will have placed the felt pieces in no particular order on the felt board.) We are going to learn about different kinds of shapes today. I am going to give each of you a piece of yarn to make shapes. First we will talk about the shapes then you will make them with your yarn.

Circle: (Hold one up) Discuss what makes it a circle. Make a circle with yarn.

Square: (Hold one up) Discuss why it is a square. Make it with yarn.

Triangle: (Hold one up) Discuss why it is a triangle. Make it with yarn.

Rectangle: (Hold one up) Discuss why it is a rectangle. Make it with yarn.

Exploring

Now that all of us have made the circle, square, triangle and rectangle with our yarn let's see what we can do with the felt pieces on the board. (Each child when called on will come up and take down the required pieces.)

Have a student do one of the following:

> All the small circles
> All the small triangles
> All the small squares
> All the small rectangles
> All the large circles
> All the large squares
> All the large triangles
> All the large rectangles
> All the medium circles
> All the medium squares
> All the medium triangles
> All the medium rectangles

Summary

Now we have talked about the different shapes, we have made them with the yarn and we have arranged the felt pieces according to their shapes and sizes, let's make the shapes one more time with our yarn. Let's make a circle. square, triangle, rectangle.

The students should be able to create the shapes with their yarn correctly.

Assessment

Hidden Assessment: Have two children at a time work together as you serruptiously assess them. Show them five designs from the classwork. The one who is "it" chooses a design from the five and gives hints to help her partner guess which design it is. The hints include the names of the shapes and their relative position in the design. Then the pair switches roles. You assess the one who is "it."

The Surprising Circle

by Baljinder Shergill-Dillon

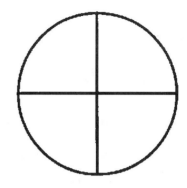

Circle Quartering

Fold and cut circles into quarters. Construct patterns using quarrters. Guess how a circle pattern was made.

Materials

Some circular objects such as plastic lids; construction paper of three different colors, scissors, crayons, and glue.

Introduction

Show several repeating patterns made from circle quarters taped to the front board and explain to the children how the patterns were made. Show how to trace a round object, then cut the circles from colored paper, fold circles into quarters and cut on folds. Working in pairs, one child traces circles, the other child cuts them out. Then each cuts circles into quarters and makes a pattern. Check for understanding.

Exploring

Pass out the paper (different colors of construction paper) for a very basic pattern.

The students will cut out circular shapes from the three colors of construction paper. After this, students will fold the circular shapes in fourths and cut on the fold. Students will make a pattern by first copying the one that is displayed on the chalk board.

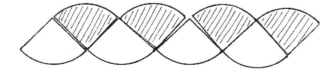

Students will make their own repeating patterns individually using any creative ideas they might have. Ask how many circle quarters they used in their pattern.

Point out how the corner of their writing paper is the same size as the corner of the circle quarter.

Summarizing

Students will show their decisions using circle quarters and will describe how they made them. Who has a pattern like this one? Who has a different pattern?

Assessment

Journal entry: When you cut your circle into quarters, you found that you got a square corner in each of the four quarter pieces. This square corner is called a "right angle." Draw a picture of your design and then describe it so that someone else could identify your design from all the rest. Describe the design in two ways: 1. What it looks like and 2. What the right angles are near or touching. Use three sentences or less.

Source: "The Surprising Circle". Marcia Danna, *Arithmetic Teacher*. January 1978.

Geometry

Alphabet Symmetry

by Julia Serences

Finding lines of symmetry

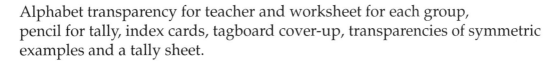

Supplies

Alphabet transparency for teacher and worksheet for each group, pencil for tally, index cards, tagboard cover-up, transparencies of symmetric examples and a tally sheet.

Introduction

We see examples of symmetry all around us. Today we are going to search for symmetry in the alphabet. Can anyone tell us what symmetry is or give us an example? Show overhead of the Westinghouse trademark. Does this give you an idea? What if we covered up half of the trademark, does the uncovered side look just like the other side? (Move cover back and forth.) Each half is identical to the other. We call such a figure symmetrical. One side is the same as the other. The line that cuts the "W" in half is called the line of symmetry. (Write symmetry on the overhead projector). Show the figure of a cross and odd shaped figure. Which of these figures is symmetrical? Discuss. Draw a circle on the overhead. Demonstrate two lines of symmetry.

Show the numbers 0-9 on the overhead. Ask the class to identify the numbers that are symmetric. Share or discuss with a partner. Ask for volunteers to come up and draw the line of symmetry for 0, 3, 8. For 0 and 8 show two lines of symmetry.

Now we are going to make a search in our groups for the letters of the alphabet that are symmetric. Each group will be given a sheet with the alphabet printed in capital letters, another sheet of paper to record letters that have symmetry and to show the line or lines of symmetry, and an index card to help test for symmetry. Make sure everyone in your group can demonstrate that the letter is symmetrical before you write it down and count it. When we are done, your group will be asked to write a symmetric letter on the blackboard and to explain to the class why the letter is symmetric and to draw the line of symmetry.

Are there any questions? Let's practice our "12-inch" (quiet) voices before we break into groups. In your 12-inch voice tell your partner how many letters out of the 26 in the alphabet you think are symmetric.

Exploring

Observe groups.

Offer assistance when needed. Are groups testing for symmetry? Are they looking for two or more lines of symmetry?

An extension for groups that finish early would be to give them some words to find the line of symmetry. Can they think of others? If the students are familiar with palindromes, ask them to find geometrically symmetric ones, i.e., 8008. Is there a pattern?

                            M
      bid   O
                            M

Summarizing

On overhead have a transparency of the letters and write in two columns headed "Is the letter symmetrical?" and "Number of lines of Symmetry."

Ask for a quiet hand from those who think that A is symmetric. Choose a group to come to the blackboard. Tell them to write the letter and draw the line or lines of symmetry. Ask if class agrees. Move on. If there is disagreement have students use a tagboard cover to demonstrate symmetry.

A, B, C, D, E, H, I M, O, T, U, V, W, X, Y all have one or more lines of symmetry. (N, S have a different kind of symmetry called rotational symmetry. (See Rotational Symmetry.) "Understanding symmetry and being able to see it helps us visually organize our environment. We see symmetry all around us." Ask students to look for examples of symmetry at home in their yards or in magazines, as a homework assignment. Make a poster or bulletin board of the things the students find.

Assessment

Have student draw two designs: one with one line of symmetry and a second with more than one line of symmetry and draw the lines of symmetry.

Sources: "Symmetry the Trademark Way" by Barbara S. Renshaw, *Arithmetic Teacher*, Sept. 1986

Family Math, Stenmark, Thompson, and Cossey. U.C. Printing Dept., Lawrence Hall of Science, 1986

WORKSHEET

A	B	C	D
E	F	G	H
I	J	K	L
M	N	O	P
Q	R	S	T
U	V	W	X
Y	Z		

0	1	2	3
4	5	6	7
8	9		

Open and Closed Paths

by Karen Harrington

The children will identify open and closed paths in their line drawings.

Materials

- Small toy animals
- Alphabet letters
- Numbers 0-9
- Pencils
- Paper
- Crayons

Introduction

- Put children in pairs.

- Draw an open path on the board. Put an animal magnet inside the open path and ask if the animal can get out. The children should respond with "yes". Identify this as an open path. Repeat this procedure with the closed path. Tell the children they will be looking for more open and closed paths in their groups.

- Familiarize children with rules of behavior. As you work you must have respect and consideration for others. You must be willing to help anyone in your group who asks. You may ask the teacher for help only if both of you have the same question.

- Model with two children the jobs each will do, e.g. one chooses a letter and gives it to the other to "test" whether it is open or closed.

Exploration

- Give each group a set of alphabet letters and a small animal.

Geometry

They will then decide which letters have open paths, which have closed paths, which have closed paths, and which have both open and closed paths. If they finish with this quickly, lead them into sorting the letters into three groups according to their type of path. The teacher circulates among the groups.

- If there is sufficient time, give each group a set of numbers from zero to nine. They will decide which kind or kinds of paths each has.

- Sometime midway have them switch jobs so each gets to be a "tester."

Summary and assessment

- Bring the students back together and make a chart on the board of the groups' findings concerning the letters (if the numbers were also used, make a chart for those, too). If there are any discrepancies between the groups discuss them at this time.

- Each child will then make a line drawing with open and closed paths, and color inside each closed path with a different color crayon.

Sorting Shapes

by Arleen R. Brienza

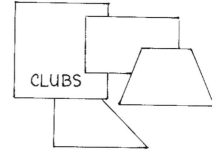

This lesson is based on the recommendation of William Burger that children need to study variety in a patiicular shape in order to learn its relevant characteristics. This activity will be done in groups of four where pupils organize shapes into "clubs"—sorting the various shapes by common characteristics. The greater part of the lesson is a group directed, active, hands-on experience.

Materials

- Ten pieces of colored construction
- White poster board paper, each with a different geometrical shape for pupils to cut out
- Glue
- Scissors
- Rulers

Introduction

Each group is given one white poster board, ten different geometric shapes to cut, glue and scissors. Without giving names to the various shapes, students are asked to cut out and sort the shapes by common characteristics into "clubs". These clubs (figures of four sides, three sides, equal angles, right angles, two pairs of equal sides, etc.) are listed on the board and students are given a sample diagram to show how to list clubs. Students are to decide amongst themselves how they will glue and record answers.

Exploration

Working as a team, students first cut, then mount the 10 shapes as indicated on the sample graph. Then collectively, they decide which characteristics each shape has, and indicate their choices by marking the appropriate box. One column is purposely left blank for the students to "discover" another characteristic. (Ask students to do this only if they have successfully "mastered" the lesson to this point.) The teacher monitors progress by observation.

Geometry

Summary

To conclude, the group discusses their findings and names the winner (the shape belonging to the most clubs). The teacher reviews this with the students and then gives students the names of each shape with a simple definition. In this, students should be able to see that different shapes can have common names and similar characteristics. Hopefully, students will recognize that squares are also rectangles and that rectangles are also parallelograms, and will be able to note differences as well as similarities.

Assessment

Give student grid paper and ask her to draw a rectangle and then explain as briefly as she can in writing how she determines if a shape is a rectangle.

CLUBS

	4 sides	3 sides	2 pair parallel sides	contains 1 or more right angles	sides of equal length

Geometry

Triangle Designs

by Pat Marshall

Developing angle concepts

Given any shape triangles, students will identify congruent, acute, obtuse, and right angles.

Materials

Overhead projector, one transparency (JOBS), felt pen, 8-1/2 x 11 typing paper, scissors, rulers, glue and paper scrap, lined paper, colored paper.

Anticipatory set

Show the class a geode. Explain the etymology of the word: ge = earth, eidos = form. Compare to the word geometry: ge= earth, metron = measure. Today we will have a beginner's study of geometry. Show inside the geode pointing out the triangular shape of the faces of the quartz crystals. Geometry is the study of common shapes on and in the earth. We will begin today to explore the triangle.

Introduction

(Have groups of four already formed.) Each group of four will make a triangle design. If your group finishes ahead of time, you may make another design and make rubbing of your designs. There are four jobs (see overhead JOBS). Let them decide who does what according to age: the youngest is the DRAWER and the oldest is the WRITER. Demonstrate each of the jobs while showing the job descriptions on the overhead. Use words: CONGRUENT, ACUTE, OBTUSE, RIGHT ANGLE, e.g. acute means "sharp" as "acute pain" and obtuse means "dull" as in "that professor's lectures are obtuse." "Your group's job is to make a unique design and then describe it so well using four or less sentences that anyone will be able to pick out the design from your description. The class will vote which design they think each description fits." Check for understanding of the job and rules for the design.

Extension

Show how to make a rubbing of the triangle design using the side of a crayon. They may do this if they finish early.

Exploring

The teacher's job now is to supervise the gathering of materials and make sure each job is being done in the groups. Notice what the children are discussing which might be worthwhile to mention during summarizing. Put the number 1-8 on the board and have each finished design posted under each number. Collect descriptions.

Summarizing

Read each description and have class vote which design they think it describes. At this time the children will be making judgments using the words congruent, acute, obtuse, and right angle. Record each vote under each design informing them if, indeed, their guesses are correct. Comment on the rubbings. Ask them how well the jobs went in each group. Tell them that when they study geometry again they may study triangles or other shapes such as rectangles or circles!

Assessment

Choose a triangle design with "everything" in it and ask student to describe everything they see.

Geometry

JOBS

DRAWER
Get a ruler, pencil and sheet of typing paper.
Fold the typing paper into thirds.
Draw a UNIQUE triangle on the top section.

CUTTER
Get scissors. From the folded typing paper, cut three congruent triangles.

GLUER
Get glue and colored paper. Glue the three triangles on the colored paper making a design.

WRITER
Get writing paper and pencil. Write a description of the design using two, three, or four sentences. Answer these questions:

> 1. What Angles were put together?
> (acute? obtuse? right? congruent? not congruent?)
> 2. What does the design look like?

When your design and sentences are finished, each group member should sign both papers. Tape the design to the front board and give the sentences to the teacher.

Rules for the design:

*AT LEAST TWO ANGLES MUST TOUCH EACH OTHER.
NONE OF THE TRIANGLES MAY OVERLAP.*

Similarity and Congruence

by Anita Laca, Cheryl Walter, Jim Cordellos

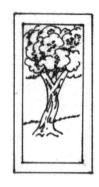

Students will identify SIMILAR and CONGRUENT shapes.

Students will learn that many geometric shapes are found in our everyday world.

Materials

3" x 1/2" and 5" x 1/2" strips of tag-board for each group of four, glue, 6x6 inch rectangles of tag-board, triangle, circle, and square shapes cut in different proportions to demonstrate similarity and congruence, actual pictures in frames.

Classroom environment

Class divided into groups of four. Children will be able to discuss and work together buildings their picture frames.

Anticipatory set

Show the pictures that you brought in.
Discuss the different frames. Are they different shapes? Different sizes?
Have students imagine a picture of a landscape. They are going to make a picture with a square or rectangular frame of yarn around it.
Assign groups of four by counting off or by more specific methods if you want certain children to work together. Then ask for volunteers to be the FRAMER and the DOER. The FRAMER distributes each length of yarn to each group of four students. The DOER gathers and distributes glue and tag-board to each student. Now it is time to make the frames.

Exploration

Students work within their group of four for the entire activity.
Each student builds his own frame made of yarn.

Geometry

Students draw their picture of the landscape they imagined within the frame.
The children compare and share their frames with their group.

Questions
Do they have all the frames possible? (5x5, 3x3, 5x3, 3x5)
Do they have repeats within the group?
Did they make a 3x5 and a 5x3? Are they the same shape? Vote on the decision.
Introduce the word CONGRUENT. These shapes are congruent. Define
CONGRUENT figures as those that are exactly alike in both size and shape.
Are the shapes 3x3 the same as the shapes 5x5? One is bigger but the shape is the same. Tape a big square to the wall and place small one on the overhead projector. Move the projector closer to the wall until the image is exactly the same size as the big square. Introduce the word SIMILAR. These shapes are similar. Define SIMILAR figures as those that are the same shape but can be different sizes.
Have the group brainstorm objects that are similar in our environment. (All squares are similar, all circles are similar. Are all triangles similar? No. All equilateral triangles are similar.)
Have the group brainstorm objects that are congruent.

Closure

Students share their pictures, first all of the squares, 3x3, and 5x5, then the rectangular pictures.

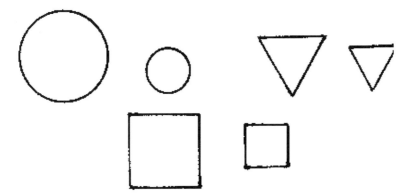

The groups show the rectangular frames with these measurements.
Come together as a whole class and make a master list of all of the objects they can think of that are similar and congruent. Observe other objects throughout the day that are similar and congruent.

Evaluation

Can all students create a frame?
Has the group come up with all of the possible frames?
Did everyone participate in the discussion?
Do they have a firm concept of the terms introduced?
Can they explain the concepts to you using your props or their frames?

Assessment

Journal entry: Students state true or false for each of these two questions and then give reasons and draw examples to prove their case.

    All squares are similar.
    All rectangles are similar.

Geometry

Rotational Symmetry

by Ortencia A. Wiley

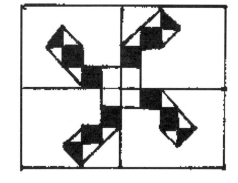

Making a symmetric design

Students will make a pattern and rotate the pattern to make a rotational symmetry design.

Divergent thinking—Students complete an original design using symmetry

Following directions—Students follow directions in tracing, patterning and pasting.

Basic skills—Student will cut squares into triangles when needed to create design.

Materials

Inch square grid (appendix) run off on different colored construction paper, white art paper, metric rulers, scissors, tracing paper, pencils, glue.

Introduction

• Review previous lesson: concentric circles, diameter and radius. Tell students have already created geometric designs with concentric circles and a compass.) Then Tell students, "Today we are going to make a design using squares and triangles."

• Ask students, "If we cut a square diagonally what do we get? Wait for answers and then demonstrate.

• Show students the designs they will use and how to make a completed design.

Show students the designs they will use...

First fold a piece of construction paper into four sections with folds crossing in the center of the paper, or measure and mark very lightly into quarters.

Paste a simple design in one corner of the paper. The idea is to rotate the design about the center point into each of the other three sections.

To help students with this idea, have them place a piece of tracing paper or white writing paper over the initial design, with the corner at the center and with two edges against the fold lines.

See figure. With the paper in this position, make a rough tracing of the design. Holding the corner of the tracing paper at the center of the construction paper, rotate the tracing to an adjacent section of the paper. This movement will indicate where the design should be pasted in that section.

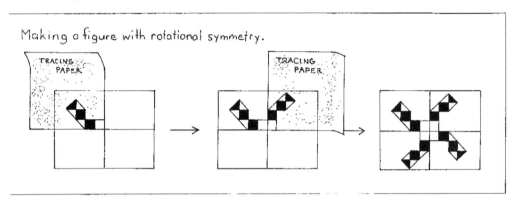

Slide the tracing to the side for reference while this new section of the design is pasted. Then rotate the design to the next section. (With four sections the design is said to have a rotational symmetry of order 4).

Tell students that each section is 1/4. One section equals 1/4, two sections 1/2, three sections 3/4, and four sections equals the whole design.

Put examples of completed designs on bulletin board for students to observe as they make their pattern for a design.

Demonstrate bilateral symmetry emphasizing that the design is flipped over rather than rotated.

Exploring

Have students make a small design using "one inch" square graph paper. Then make a design as shown.

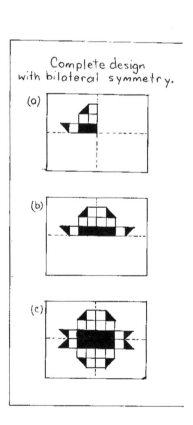

Geometry

Summarizing

As part of their summarizing, have them show you how they got the design. Then have them place their signed designs on a bulletin board display.

Assessment

Have students choose two student-made designs - one with rotational symmetry and the other with line symmetry. Ask which is which and how she determines if a design has rotational symmetry.

Source: *Arithmetic Teacher*, September, 1984, pp. 8 - 13.

Polyhedrons

by Suzanne T. Tallman

The students will each construct one of the following polyhedrons while becoming aware of their names.

Anticipatory Set

Bring out three paper bags, each containing an object in one in the three shapes., i.e., dodecahedron - soccer ball; tetrahedron - pyramid; cube - dice or box. Tell the class that in each bag you have objects that have specific names for their shapes. As I pull the object out, take a look at it and tell me what shapes you see. Show one object at a time and give students "think time." Then tell the class the name of the three dimensional object, writing it on the board for them to see. Go through the pronunciation and have the class repeat.

We are going to be making these shapes. Each of you is going to make one of these shapes.

Purpose

The student will receive hands-on experience with three-dimensional geometric objects.

Materials

- Geometric shape dittos (attached)
- Physical example of each shape
Soccer ball (Dodecahedron)
Pyramid (Tetrahedron)
Die or box (Cube)
- Glue or paste
- Scissors
- String
- Three paper bags

Geometry

(Icosahedron)

INPUT	MODEL	CHECK FOR UNDERSTANDING
As I pull the object out, take a look at it and tell me what shapes you see.	Show one object at a time.	Children volunteer the names of the two-dimensional geometric shapes that make each shape.
Tell class the name of the three-dimensional shapes.	Write name on the board.	Children chorally pronounce the name.
You are each going to make one of these polyhedrons.	Show the handouts.	

Guided Practice

Assemble the icosahedron while verbally describing the cutting, folding, and gluing (or taping) involved.

Independent/Group Practice

Children independently construct a shape. Have finished models in a central location in the room for children to consult as they assemble theirs.

Closure

Tape string to the shapes and hang them around the room.

Assessment

Pose this problem using a net of six squares ☐☐☐☐☐☐
Pat found she could not make a cube by folding and taping this net. Help her draw a different net that would actually make a cube. Make it different from the one you already made.

Polyhedrons

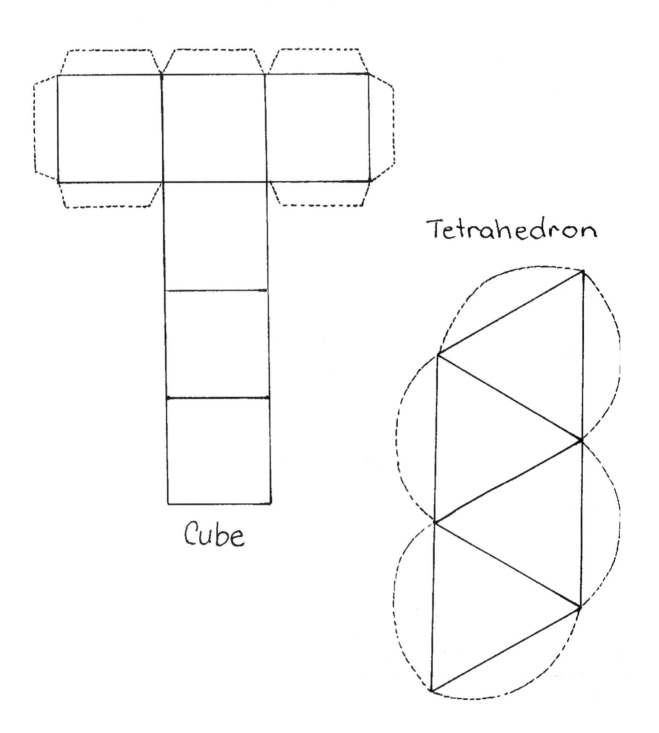

Cube

Tetrahedron

Geometry

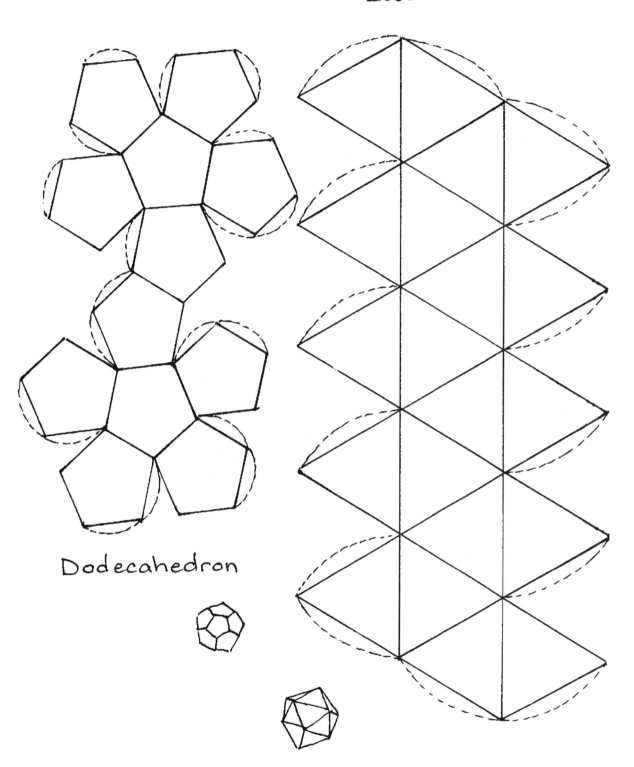

Battleship Game

by Yvonne B. Manley

Graphing Ordered Pairs

Given the coordinates, students will graph a point.
Students will correctly read out the coordinates for a given point.

Materials

- Overhead projector or chalkboard for examples
- A sheet of centimeter graph paper for each student
- Pencils

Introduction

Teacher asks how many of the students have played the game Battleship before. She then elicits the procedures for playing the game making sure they discuss how they locate their opponent's ships by calling out the number of spaces across and up. She then substitutes the words "column" and "row" for "across" and "up." Show the symbolism (1,2) to represent "across 1 and up 2." To make the students "buy" into the lesson she tells them that they are going to make their own board and that at the end of the lesson they will play a game of Battleship using their boards. The teacher can then talk about how a similar graph system is used to locate and track planes and why such a system is used.

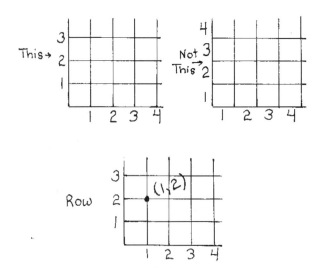

Geometry

Procedures

On the blackboard draw a graph with the axes labeled. (Note that the lines are labeled, not the spaces.) Then plot a point and explain how to label it. The number of the column is read first and then the number of the row.

Have a monitor distribute graph paper to each student when everyone is ready, name the coordinates of various points for the students to plot. A good motivational device would be to call out points that form some type of design when connected. For instance, (1,2) (1,8) (7,13) (12,8) (12,2) (9,2) (9,4) (7,4) (7,2) (1,2) makes a house.

Then draw the Battleships on the graph and explain the rules of the game. To start, each person draws three ships on his or her grid, hidden from each other. The ship must be seven squares big.

Rules of the game

Each player will take turns guessing the coordinates of the other player's battleship.

If the guess is correct, the player says, "Hit". If incorrect, the player says, "Miss".

If the player guesses correctly, she/he can guess again.

Each student will keep a record of the points she/he has called. This list can then be used to verify the guesses should a disagreement occur.

Play continues until one player has sunk all of his or her opponent's battleships. It takes three hits to sink a battleship.

Model one game with a student. The students then break up into pairs to play their game.

Assessment

Using a globe or map, ask student to write the latitude and longitude of a place in the world they would like to see someday. Let the class locate the place on their maps.

Source: Smith, Robert F., "Coordinate Geometry for Third Graders." *Arithmetic Teacher*.

Circles

by Joy Peacock

The students will explore the concepts of radius and diameter by arranging themselves to form the radius and diameter lines with the dodge ball circle. They will discover the relationship between diameter and circumference by measuring round things, charting results and drawing conclusions. They will form and test hypotheses about the circumference of the dodge ball circle in this circle game.

Materials

Outside: A dodge ball circle with a dot drawn in the center, sidewalk chalk, 12 inch string, measuring stick, measuring tapes, grid paper, pens or pencils, chalk, chalkboard, round things: cup, bowl, plate, towel roll, cans, etc.

Introduction

Hold up the round things and ask what they have in common. Explain that we will learn how to say exactly how big a circle is in a mathematical way.

Hand each student a twelve-inch string or measuring tape to practice measuring skills. Have them measure their desks, their book, etc. just to make sure they understand the process of measuring.

Procedure

Take two groups of four to the dodge ball circle. Form two lines from the center to the edge of the circle. Mark the center D and the points on the circle A and B. Explain that we call this circle D because that is what I named the center point. The lines are called DA and DB. Have the children make new lines from one end of the circle to the other and intersecting the center point. Use all eight children. Label these points and name the line. Explain that this is called "diameter". See if all lines of diameter are equal in the circle by moving the line. Try not going through the center to see the changes. Have a child step out of the circle so the line doesn't touch one side and ask if that is a diameter line. Have students explain that the line must touch both sides of the circle and go through the center.

Geometry

Go inside and into groups of four. Measure the round things using either string/measuring stick or measuring tape, whichever they prefer. Demonstrate one measurement and how to record by listing the object, the distance around (circumference) and the distance across (diameter). After all have measured and recorded several, have them look for a relationship to see if they find about 3:1.

Object	Diameter	Circumference

The groups share their findings using the new terms "diameter" and "circumference". Place on board. Repeat with three other groups of four.

Closure

Whole class activity—Propose this question and write on board: "If the relationship between diameter and circumference is about 1 to 3 and it took 8 students to go across the circle, how many students will it take to go around the circle?

Students discuss in their groups and share answers.

At next recess have the whole class go outside to test their hypotheses.

Assessment

Give student a string to measure the circumference of her partner's head. Ask her to predict the approximate diameter of her head and explain why she thinks her answer is right.

Projective Geometry

by Margaret-Ann Vansoest

Shadows change shape depending upon how the projected object is moved.

- Students will observe through exploration that the relative position of an object to its light source will change the shape of the shadow that is being projected.

- Students will explore hand shadows made from a point source of light and the effect that distance has on the form of the shadows.

- Students will record examples of the above observations.

- Students will explore and record observations made from the projected shadow from a square piece of cardboard.

- Students will discuss findings of their group with class.

Materials

Blackboard, chalk, one flash light for each group of four, a paper projection screen for each group, paper and pencils or felt pens for recordings of each group, cardboard shapes (square, equilateral triangle and miscellaneous shapes) for each group and a sheet for each student to write the general findings.

Introducing

I would review the meaning of the terms "obtuse, parallel lines, congruent, similar, square, right angle, acute and obtuse angles, polygon, and triangle" with the students. I will tell them that the size or shape of an object's shadow can be changed depending on its distance from the light source or the "tilt" of the object. "I would like you to explore the kinds of shapes that are projected upon the wall when you use your hands and a flashlight. See how moving your hands closer or further from the light source affects the shadow. Also note how tilting or moving changes the shape of your shadow. After you write down your observations in your groups I will check what you have done on paper before you explore with some shapes that I will give you to work with. Who can tell me what it is that each group will be doing so that I can write down your steps on the board as a reminder?"

Students will answer that they will observe what happens to the shadow when the object is tilted or moved away from the light and record what you observe.

Write on board:	a) Effect of distance
b) Effect of tilting

Demonstrate the set-up of the light source (flash light), hand shadow and wall. Demonstrate how observations are recorded, using the vocabulary words, e.g., "The shadow of the triangle is similar to the cardboard triangle. When we moved the cardboard away from the light the shadow got smaller but stayed the same shape."

List on the board vocabulary words reviewed earlier as a reference for students: congruent, parallel, lines, similar, angles, polygon, triangle, square, right angle, obtuse angle, acute angle.

Assign jobs to be handled in each group: flashlight person, shadow making person, recorder, coordinator.

Exploring

Have the group coordinators gather the materials for their groups. Check individual groups for their interactions, set-up, and participation in the activity. Answer questions that may arise amongst the students. When a group is finished with hand shadows, check their recordings and give them a square cardboard shape to explore and record findings. If time allows, give groups triangular shapes or other geometrical shapes with the same set of instructions. Perhaps have the students come up with their own shape to explore if they finish with the other shapes.

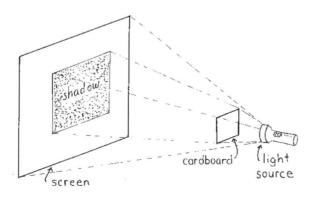

Summarizing

Have groups report back to class on their findings. Have each group describe one observation or finding and then go on to next group. Ask them not to repeat what another group described. Teacher lists on board. Ask if all students agree with each of the findings. Have each student record the general findings on a sheet titled "Projective Geometry": 1. Hand figure, 2. Square, and 3. Triangle. Challenge students to explore in their free time in class or at home and report their findings back to the class.

Assessment

Open ended: What did you see when you moved your square in different ways in front of the flashlight?

Close-ended: "When your square moved close to the flashlight, what happened to its shadow? When you tilted your square, what happened to its shadow?

Geometry

Triangle Inequality and Triangle Nests

by Rebecca B. Fuentes

Properties of triangles

Develop the concepts of triangle inequality and triangle nests. Connect between the concrete and symbolic representations of these mathematical concepts.

Students will construct triangles, record measurements, form triangle nest, record corresponding sides, write in.

Materials

Paper for work space, paper clips, strips of grid paper with lengths of 3, 4, 5, 6, 8, 10, 12, 16, 20, and 24 units, tape, lined paper, pencils, sample triangle nest, a picture of a geodesic dome.

Anticipatory set

(Show a picture of a geodesic dome. Describe it.) This was invented by Buckminster Fuller. It is a very sturdy structure made of triangles. It is so sturdy because triangles are a unique shape among all the polygons. You will make your own triangles and find out what is so special about them.

Procedures

Each group of four will make 10 triangles. Form groups of four and describe jobs: DOER gets materials, FORMERS make triangles from paper clips or paper strips, RECORDERS write down the dimensions of each triangle made by the group.

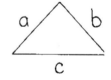

Hand out manipulatives (approximately 20 paper clips per student and one of each length of paper) and have them construct five triangles with paper clips, and five with paper strips. Tell them to record the lengths of the sides of the triangles on the paper. Show them the form: "3, 5, 4" for a triangle with sides of these lengths. If they are not shown this form, many students will add the sides together and simply write down "12". When they have four to six triangles, go to front board, draw a triangle, label it (see above), and introduce the equation $a+b>c$.

Explain what it tells us about triangles, using some of the students' triangles to show it physically. Put two sides together and show that, together, they are longer than the third side. Then show that we can also represent this with numbers. Count how many paper clips or inches make up each side, write that down and then the length of the third side. 3+4>5, 4+5>3, 3+5>4. Have them try the equation with the rest of their triangles. When they are finished, introduce triangle nests. These are similar triangles "nested" inside one another (see drawing below.) Show them your sample and tell them the lengths of each side. Ask if anyone sees the pattern. Have students who think they have it try to explain it for the students who did not catch it. Have students construct nests. (Don't be surprised if someone asks for some mud and straw.)

Summary

After everyone has constructed a nest, have them clean up and gather the manipulatives. Ask them if three strips that are 3, 5 and 9 units long would make a triangle. (No) Ask them to write you three combinations of numbers which will make triangles, as well as three triangles which form a nest. For the nest, ask for a drawing which should be labelled. Ask them what triangles formed nests. (Record on board.)

                3, 4, 5

                6, 8, 10   "What patterns do you see?"

                12, 16, 20

Triangle inequality theorem: Any two sides of a triangle, when added together, are longer than the third side.

Triangle nests: Triangles of the same shape but different sizes have corresponding sides which are multiples of each other.

Assessment

Open ended: What did you learn about triangles from making these nests?

Geometry

Fun and Folklore with Tangrams

by Sara Fries

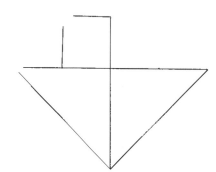

Make a tangram and tangram design

The student will:

- identify geometric shapes (triangle, square, rectangle, trapezoid, parallelogram)
- be engaged in tasks which foster spatial visualization and problem solving abilities
- listen carefully so as to be able to follow directions on how to cut out a Tangram

Materials (for each student)

1 square piece of construction paper (e.g., 10cm. x 10cm.), 1 pair scissors, 1 piece heavy paper or tagboard (8 1/2" x 11"), 1 piece chart paper (16" x 24" approximately), pencil, marking pen, ruler (optional)

Procedure

1. Give students background on Tangram: The tangram had its origin in China. The legend tells us that a man by the name of Tan lived in China about 4,000 years ago. His most prized possession was a handmade ceramic tile. While carrying it to the palace to give it to the emperor, he dropped the tile and it broke into seven pieces. He spent many years trying to put the tile together again. To the delight of his friends, he was able to make more than 300 designs with the seven pieces. Tan's puzzle was passed on from one generation to the next and is still a favorite puzzle today.

The Tangram consists of seven basic pieces:

- 2 small triangles
- 1 medium triangle
- 2 large triangles
- 1 square
- 1 parallelogram

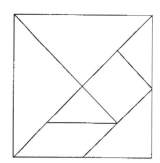

2. Review shapes (triangle, square, rectangle, trapezoid, parallelogram)
3. Elicit definitions of shapes from students
4. Pass out construction paper squares, scissors
5. Instructions to cut out the pieces are on the next page.
6. Those who finish quickly can be challenged to all the pieces back into a square.

Exploring

1. Have students use just the three smallest triangles to make a square. Then use those same pieces to make a triangle, a rectangle, a trapezoid, a parallelogram. Then use the five smallest pieces to make the same shapes. Repeat with all seven pieces. They must trace around the shapes they use, and they must label the shapes they are

TANGRAMS	square	triangle	rectangle	trapezoid	parallelogram
3 small triangles	◨				
5 small pieces					
all 7 pieces					

trying to make. Have them record on a chart as show:

2. Tangram Puzzle Cards. Have students explore making shapes using all seven pieces. When they find ones that please them, have them draw around the outline of the shape on heavy paper or tagboard. Then name it, sign it, and put it in a class Tangram box so others can try to fit their pieces into the shape. Students sign their names on the backs of each others' puzzles as they solve them.

Assessment

Before assessing the student, make two copies of the shape he made for the Tangram Puzzle Card. Rotate one copy and flip the other and glue them each to a piece of paper. Then ask the student to come in the room. Tell him what you did to the shape (rotate and flip) and ask him if he thinks he can figure out whose Tangram Puzzle Card you did this to. Then show him the two copies. (It takes some spatial ability to visualize his shape in a different orientation so some students will not recognize their own shape.)

Making a Tangram

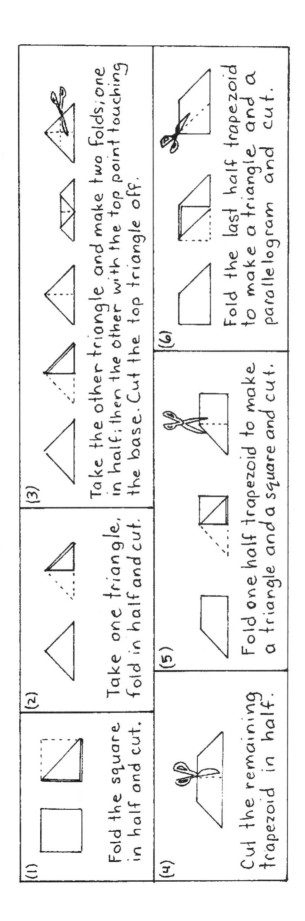

TANGRAMS

Name: _____

Can you make → with this many pieces? ↓	Square ☐	Rectangle ▭	Triangle ◣	Trapezoid ▱	Parallelogram ▱
1					
2					
3					
4					
5					
6					
7					

Tessellations

by Patricia Welty

This activity involves finding tessellating patterns by using different shaped polygons. Also, this activity will allow the students to work in groups of four or five.

Materials

Each group will need an envelope with eight polygons cut out of construction paper. Two triangles (right-angled and obtuse-angled), four quadrilaterals (kite, parallelogram, rectangle, trapezoid), two other polygons (hexagon and octagon). They will also need several sheets of blank paper.

Introducing

Anticipatory set

Have you ever heard of the word tessellations before? Tessellations are tiling designs that cover a flat surface without leaving gaps and without overlapping. You have probably seen tessellations in your kitchen or in your bathroom.

Pose a Part of the Problem

Here is an example of one and of what you will be doing later in your groups. I am going to take a rectangle and show you four different tessellations that I can create.

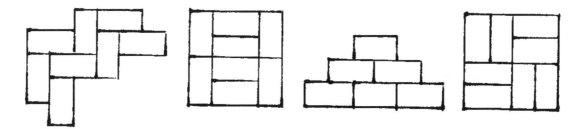

Show on overhead projector.

Then show them a pentagon that doesn't work, and explain why. Also, show some examples of some that are wrong because they overlapped or have spaces and ask them to tell you why they are wrong.

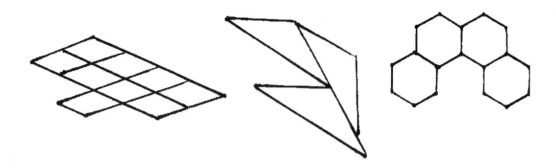

Present the Problem to be Solved

Here is the problem that I want everyone to help me solve today; Oftentimes the only tessellations that you will see will be patterns made with square. I am going to give each group an envelope with eight polygons (show). I want you to use one shape at a time and see how many different patterns you can come up with for each shape. You will do this by choosing one and tracing it over and over until a pattern does or does not appear. There are two rules that you need to remember (write on overhead) 1) The tiles should not overlap 2) No spaces should be left between the tile. Before you hand out the materials number the members of the groups 1, 2, 3, 4. Then write down the jobs one at a time for each group member, and tell them that they will change jobs every 6-8 minutes. You will say, "Switch," and the 1 will become 2, etc. Here are the jobs:

1) Doer: gets the materials, counts them, and turns them in when the groups are done.
2) Manipulator: the one who moves the objects around.
3) Recorder: traces the shapes.
4) Quality Controller: checks the manipulator's work to make sure that the rules are followed.

Discussion

"Before I hand out the materials, are there any questions?
Okay, number ones please come up."

Exploring

Circulate and observe. Encourage them to help each other with ideas.
Make sure that they are all doing something and not just watching.

Geometry

Extension

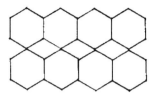

Have them create patterns using two or more of the polygons. Show them a book of art by the artist Escher.

Summarizing

Go back to the original problem and ask it again. Ask different groups to share their patterns and also have them share a few that didn't work. If some of the groups did the extension, have them share their results.

Assessment

Have students use the shapes in the envelope to show you the answer to the question: Which shapes formed a tesselation and which did not?

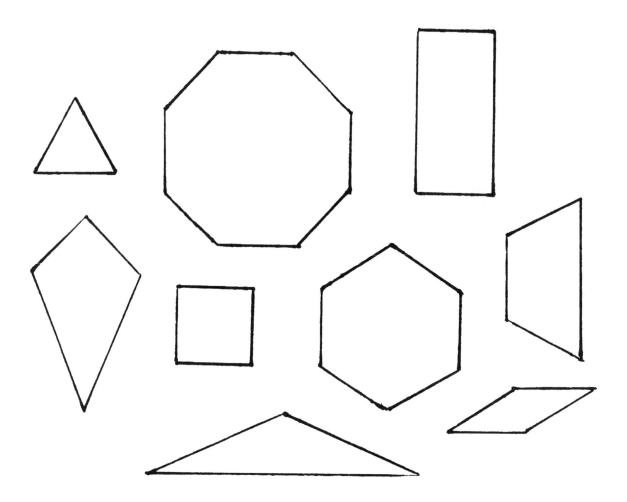

Tessellation Patterns

Geometric Quilt Bulletin Board

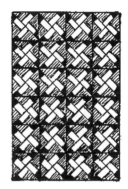

by Judith Nelson-Ullery

Students will work together in teams to produce a student made geometric quilt using a patterned design.

Materials

- Various colors of construction paper
   - Squares/rectangles, pre-cut to measure
   - Paper cut to shapes from (Note - project must be planned to account for varying bulletin board dimensions)
- Shape templates for children to trace onto paper and cut out. Square, circle, star, triangle, pentagon, etc.
- Scissors, glue, pencils, stapler (Note - Gluing patches on large piece of poster paper will allow you to move quilt. Could be used as a tablecloth for party.)

Anticipatory set

I'd like you all to help me fill this empty bulletin board. We're going to cover it in a quilt. Have any of you every seen a quilt? It's many patches of fabric stitched together to form a blanket. We'll make a quilt for our board made of paper.

Introduction

Today we will make our patches and stitch them together on the board This will be a 'geometric' quilt. I have some shapes here that you can take turns tracing onto some colored paper, then cut them out with your scissors. While you are cutting the shapes I will pass to you the colored paper 'patches'. I'd like you to glue your shape as close as you can to the center of the patch. (Model example - demonstrate by making patch, check for understanding.)

Exploring

- Have students work in teams of two.
- Have them divide job; team members alternate use of templates and scissors.
- Specify how many patches each child should make
- Pass out templates and paper to cut shapes from

- After shapes cut distribute proper color patch to each group. Have students paste shape on center of each patch.
- When all materials cleaned up have students regroup in front of board with their patches
- One by one have them come forward - alternating partners each turn - with appropriate patch
- Ask students if they can see pattern as it forms. Ask them to predict which patch will come next
- Having students tack, pin, or glue the patches on themselves frees you to monitor children better
- Check for understanding by allowing students to tell when to come up to board
- After quilt complete have students pick up a Name Pattern ditto. Read and follow directions on ditto. Have students write what they found on the back of the sheet.

Guided practice

"These are wonderful patches. This quilt is going to look beautiful. The first thing we need to do is have your teams with your patches regroup from the work area to in front of the board. We must decide in which order to place the colors. Let's begin with the red patch with the black square. Would one person from your team place one patch in the upper left corner of the board. The next time we need your color patch have your partner put in up. OK, what color should we have next? Yellow - good - will one person from your team please put up a patch". Yellow - good - will one person from your team please put up a patch". Continue. At second red patch remind team to switch partners if they had forgotten. Do this with each team until they understand taking turns. Ask students if they can see the pattern as it forms. Ask them to predict which patch will come next. Continue until board covered.

Sponge activity

See "Can You See It In a Name"

Summary

Today the students have had the experience of handling geometric shapes by tracing, cutting and gluing the. They have practiced working as a team and dividing the work load. Students discovered the forming pattern as they produced the Geometric Quilt. The idea of patterns was reinforced with the Name Pattern ditto.

Assessment

What did you like about working together to make the geometric quilt?

CHAPTER 3
Measurement

Grab Bag Comparisons   Comparing lengths   K-2
Heavier and Lighter   Comparing weights   K-3
Time Order   Putting events in order   K-1
Paper Clip Measuring   Using non-standard units for length   K-3
Ready, Set, Measure!   Using non-standard units for length   K-3
Making a Meter Tape   Cutting grid paper to make a meter tape   3-6
Round Things   Finding pi   4-8
Stringing "Em Along   Length in standard and non-standard units   4-6
Big Foot   Estimating and measuring area   4-8
Slicing Rectangles to Find Area   Area of a tringle   6-8

Grab Bag Comparisons

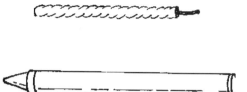

by Michele Denton Hanel

Comparing Lengths

Children will compare the length of variety of objects with other objects, such as a paperclip, a straw, or a length of paper.

Materials

- zip lock bags
- straws
- paperclips
- strips of paper (any size you want)
- objects to compare such as crayons, pencils, spoons, candles, unifix cubes, and legos
- Work boards - You can use a large piece of construction paper.

Anticipatory set

"How many of you have ever measured something before? Do you know what we use for measuring? Have you ever seen your Mom or Dad measure something before? If so, what?"

Objective/purpose

I am going to show you how to compare objects to see which is bigger, smaller, or the same. This will help familiarize you with comparing objects before you learn how to measure using standard units.

INPUT	MODEL	CHECK FOR UNDERSTANDING
"In the zip lock bag I have placed a bunch of common items like crayons, pencils, candles, spoons, unifix cubes, pegs, and legos.	Show them the individual items in the zip lock bag.	Have children raise each item in the air as you call each item out.

Measurement

INPUT	MODEL	CHECK FOR UNDERSTANDING
There is a bag in front of each of you."		
"You will also see in front of you a straw, paperclip, strip of paper, and a work board. These are the things you are going to compare with the things in the zip lock bag ."	Show them which item is the straw, paperclip, and strip of paper.	Have children point to the straw, paperclip, and strip of paper.
I would then ask the children to choose the longest measuring device (straw) and then sort the items in their zip-lock bag into 3 piles. "Each pile will consist of all the items in their zip lock bag that are longer than, same, or shorter than the measuring device (straw). Each pile will be placed on your work board. As you will see, your work-board is divided into 3 parts. One part is labeled with a long line, the middle part is labeled with two lines the same length, and the last part is labeled witha short line. Please put all the things longer than the straw on the section labeled with the long line, all the things the same length as the straw in the middle of the paper, and all the things shorter than the straw on the section labeled with the short line."	I will show the children that when they compare two items they must always make sure that both items are touching the table. Demonstrate this! 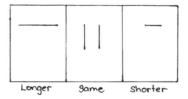	Walk around and observe each child to make sure he/she is measuring correctly. Also make sure the children are placing the items in the appropriate piles.

I would then repeat this procedure using the paperclip and strip of paper.

Practice

Guided Practice: Create your own workbook page. Go over the first two problems on the workbook page to make sure the children know what they are supposed to do. Make a workbook page that has several comparisons on it. Ask children to put an "X" on the item that is longer, same, or shorter. For example, have the children put an "X" on the tree that is the shorter of the two shown. Check for understanding before moving on to independent practice.

Independent Practice: After child is through with his/her workbook page, give child a piece of string and have the child go around the room and find an object that is longer than, same, and shorter than the piece of string. Have the children report their findings to you when they are done.

Summary

I would have the children tell the person sitting next to them what things they compared with the string.

Note: You may use any common items as nonstandard units of measurement. The ones I used were items I found around the house.

Assessment

Demonstration: Use the independent practice activity as the assessment. Each child reports to you what objects were longer, shorter and the same length as their piece of string.

Measurement

Heavier and Lighter

by Kathleen A. Sutphen

Comparing Weights

Students will demonstrate comprehension of "heavier-lighter" by correct verbal response and show of hands to the question, "Which of these objects is heavier (or lighter)?"

Materials

A balance beam
Objects to compare e.g. bagel and rice cake, large cotton wad and small lead piece, brass apple and real apple, candle and foam curler, aluminum and lead cylinders, rock and little pumpkin, two blocks of wood, quartz egg and cookie cutter.

Anticipatory set

"Who knows what it means to weigh something?" "Who knows what it means to say something is heavier than something?" "What does lighter mean?"

Objective/purpose

Today we are going to weigh some things that I brought in so that we can learn what it means to say an object is heavier or lighter than another object.

INPUT	MODEL	CHECK FOR UNDERSTANDING
Our scales will show us which of these pieces of fruit weighs more than the other. I have an apple and some grapes here. Who thinks they know which weighs more?	Apple and grape. Place one in each tray on scale and compare weight.	With their verbal response as I call on them.

INPUT	MODEL	CHECK FOR UNDERSTANDING
I will ask the children to indicate with a show of hands which fruit they think the scale indicates is the heaviest.	The scale with the fruit in it.	With a show of hands.
Two children will be selected and asked "Which of these two pieces of wood weighs more?" The children will be encouraged to handle the objects before making their predictions. (Similar block but one is larger)	The children will place the wood in the scale and the class will observe the outcome.	With correct verbal response.
The pairs of children will be asked which item is heavier as indicated by the scale. The class will be asked to show with their hands if they agree.	The scale with wood in it.	With correct verbal response and a show of hands.
The above will be repeated with sets of two children until all students have predicted weight and determined weight of their objects using the scale.	Each group of two will weigh their items after making their predictions.	Same as above.

We will not be taking actual numerical figures for weight, just "heavier & lighter."

Practice

Practice is included in the lesson with the children coming up in pairs to predict weight and to do the weighing.

Assessment

Demonstration: Hand child two objects to hold (e.g. an apple and an orange), one in each hand, and ask her which is heavier/lighter. Then ask her to predict what will happen when she puts the objects at either side of the balance beam. Student checks prediction by putting objects on balance beam.

Measurement

Time Order

by Kathy Mather Patschke

Students will put events in time order: first, next, and last.

Materials

Teacher
Pictures of pumpkin and boy scenes, 2 pieces of paper, scissors, and paste.

Students
Scissors, paste, 3 pieces of blank paper per student.

Supervising Adults
A copy of the guided practice, independent practice, extension and the summary.

Anticipatory set

I will talk about the sequence of the school day, asking the children to name specific activities that take place between the time they arrive at school and the time they go home. After the children have named three activities I will ask them to help me put these events in order. I will use the words" first, next, and last" as I help the children describe the school activities.

Objective/purpose

Show which of three events happened first, next, last. It is important to know which comes first, next, and last in an event.

INPUT	MODEL	CHECK FOR UNDERSTANDING
I will tell the children that these pictures show three parts of an activity in mixed up order.	I will show three different pictures that show the growth of a pumpkin.	Look at the children's faces.

INPUT	MODEL	CHECK FOR UNDERSTANDING
that I will now put these pictures in order and tell the children the flower grows first, next the flower becomes a little pumpkin, last the pumpkin grows big.	Put pictures in order and show the pictures of the growth of a pumpkin in order from left to right.	Look at the children's faces.
I will ask the children if these three pictures are in order.	I will focus on three pictures of a boy. These pictures will be in mixed order. I will cut out each picture.	I will have the children put one hand in the air if they think the pictures are in order. Hands down if not.
I will ask the children to raise a "quiet hand" if they can tell me which of the three pictures happens first and why.	Still focusing on the pictures of the boy. When a child selects the first picture, I will paste it on the left side of the paper.	I will ask the children to raise their hands if they agree. (I will do this until I get the correct response.)
I will ask for another volunteer to tell me which picture comes next and why.	Still focusing on the pictures of the boy. When the next picture is selected, I will paste it in the middle.	Hands up if you agree.
I will ask for another volunteer to tell me which picture comes last and why.	Still focusing on the pictures of the boy. I will paste the last picture on the right side of the paper.	Hands up if you agree.
I will tell the children that they have now put these events in order – first, next, last. I will tell the children they are going to cut some pictures of paper in order – first, next, and last, just as I did with the boy pictures.	I will hold up the sequence of pasted pictures. I will point to first, next, and last.	I will have the children point to first, next, and last.

Guided practice

I will have the green group go to the table by the bookshelf and work with (teacher aide). I will have the blue group go to the round table and work with (aide or parent). The purple group will work with me. I will have the children cut out the three pictures of a winter scene. I will then have them paste each picture on a piece of paper in order—first, next, and last. While the children are doing this the adults aiding at the table will be checking, helping, and answering any questions. The adult should look at each finished paper to make sure the events are in order. If not, help the children put them in order. (The Winter Scene is the most ambiguous sequence in this set so let the children describe why they used the order they did. Did the boy started at the top of the hill or the bottom ?)

Independent practice

After the children have successfully put the Winter Scene in order, they will work independently on the Balloon Scene.

Extension

In the blank frame they may draw what happens when they drink their favorite drink with a straw, or make up their own time sequence story.

Summary

As each student finishes the pictures, he/she will show them to the aide at the table. The children will tell the aide why they put the pictures in that particular order.

Measurement

Paperclip Measuring

by James Gillespie

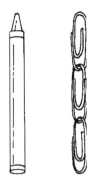

Using non-standard measures for length

Given measuring devices made of strung-together paperclips, students will determine the lengths of objects the same length or shorter than these devices accurate to within one unit.

Materials

- Overhead projector and screen
- Paperclip measuring devices (10 clips per string)
- "Paperclip Measuring" recording sheet (two per student).
- Objects for demonstration: e.g., pencil, chalkboard eraser, small book, crayon
- a small object from their desks for each student to measure during guided practice

Anticipatory set

Ask the children how we might find out how tall they are, how wide their desks are, or what size shoes they wear.

Objective/purposes

I am going to show you one way to measure length. This will help you to understand how measuring is done.

INPUT	MODEL	CHECK FOR UNDERSTANDING
Explain linear measurement. 1. Decide what to measure (a pencil). 2. Decide what units to use (paperclips). 3. Compare what you are measuring to the units. 4. Count the number of units along the measured item's length. 5. Record the measurement.	On overhead, measure a pencil, being certain to clearly show the comparison of the units to its length. Stress the lining-up of one end of the object being measured with one end of the measuring device.	Have all students count along with teacher as units are counted along pencil's length.

Measurement

INPUT	MODEL	CHECK FOR UNDERSTANDING
Describe improper measuring techniques.		
1. Not lining up. 2. Object too short for units. 3. Object too long for device.	Show improper lining-up of the object and the measuring device. Show problems of measuring a pencil with your forearm.	Have all students count along with teacher as units are counted for each incorrect technique.
Describe how to measure a different object (a chalkboard eraser) as above.		
1. Note that the same units can be used to measure different items. 2. Note that the same device can be used to measure different items. 3. Note that both length and width can be measured.	Model the correct measuring and recording procedure. Show measurement of both length and width of eraser.	Have all students count along with teacher as units are counted along eraser's length and width.
State that you are going to measure something else. "We'll start with length of this book. Show me with a silent signal if you think this length is more than the units."	Show measurement of length of a small book. Model hand signals.	All students signal agreement/disagreement with silent hand signals.
Say you will now measure the book's width. Ask for predictions.	[agreement / disagreement — Silent Hand Signals] Measure the book's width.	Each student predicts the width of the book aloud. All students count along with teacher as book's width is counted up.
What thing am I measuring? What units am I using? How many units long is it?	Show measurement of a crayon.	All students count along with teacher as crayon's length is measured.

Guided practice

Give the students paperclip measuring devices, objects to measure, and recording sheets. Teacher observes and checks students as they practice on the recording sheet they draw a picture of the object and then record its length and width.

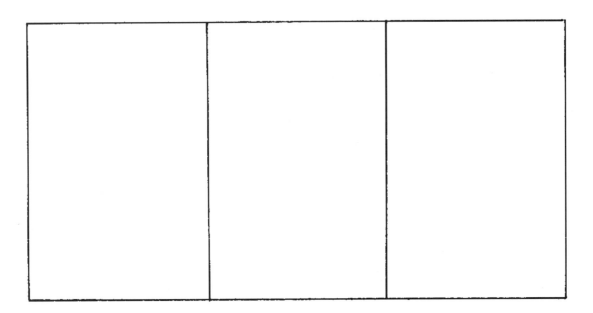

Independent practice

Have students take paperclip devices and new recording sheets home, measure the items of their choice and bring back results.

Summary

Tell me what you learned today about measurement and why it might be important to know how to measure correctly.

Assessment

Journal writing: Have one student (e.g. Winston) hold a large paper clip and another student (e.g. Yuri) hold a small one. Then pose this problem and have them write their viewpoint:
> Winston measured his pencil with his paper clip and found it was four units long. Yuri measured his pencil with his paper clip and found it also was four units long. Are their pencils the same length?

Ready, Set, Measure!

by Nancy A. Noma

This is an activity that uses non-standard measurements to introduce and explore the concept of length.

Materials

Paper clips, 4" nails, toothpicks, thumbs (supplied by children), 6" lengths of colored yarn, 18" lengths of white string.

Introducing

Present the concept of length.
Have students give examples of longer, shorter and about the same.
Guess, measure and record.
Using the white string, measure some classroom objects, such as the chalkboard, a chair, a table, etc. First, make a chart on the board (see below). Then have the students guess the measurement, and last the measurement. Use three objects.
Present rules for working together.
One person records, one person measures. Trade duties.
Present the problem to be solved.
Each team of two students will be given a bag of paperclips, a bag of nails, a bag of toothpicks, 6" lengths of colored yarn and the sheet for recording their measurements. Each team is to measure the white string using each of the non-standard units of measurement. The students are to record the results of each measurement in the box next to the picture of the unit of measurement. Point out the fact that thumbs are not all the same size. The team must work together to measure. Model this with two students. Give direction to begin.

object	estimate	actual

Exploring

Circulate and observe. Help out if teams need assistance. Observe to see if teams are working well together.

Summarizing

Discuss team processes.
Ask how they liked working in pairs. Ask if it is more comfortable to work as a team or by themselves. Ask if they worked together to measure.

Have each team fill in the chart on the board.
Use their own team chart as a reference. The board chart will show each team's recorded measurement for each of the five objects.

Pose questions.
Which objects were the hardest to use for measuring? Why? Which were the easiest? Why? Did it take more paper clips than toothpicks to measure the string? After recording the information from each group, there may be other questions to be asked, depending on the variations in answers. e.g. for which objects did all teams get within one unit of the same answer? Emphasize the idea that close is O.K. Exact measurement is impossible to achieve.

Assessment

Observation: Observe how students use the various objects as units of measure. For example, do they try to insure that units are not overlapping or leaving spaces as they measure?

Measurement

PAPERCLIPS	
NAILS	
STRING	
TOOTHPICKS	
THUMBS	

Making a Meter Tape

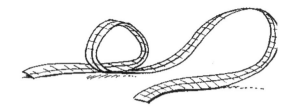

by Priscilla Huff

Concepts of meter, decimeter, and centimeter.

Given the necessary materials and instructions, students will make a metric tape measure one meter long marked off by centimeters and decimeters.

Materials

- metric graph paper
- pencil, paper clip
- scissors for each child
- tape

Anticipatory set

How tall are you? Do you know your metric height? We will each make a metric tape measure to find out!

Objective/purpose

Students will learn to use the metric units on a meter stick.

INPUT	MODEL	CHECK FOR UNDERSTANDING
Introduce metric system. Explain the features of the paper. (e.g. Grid is 15 cm X 20 cm)	Poster (metric nicknames), the grid paper. Show one sq cm by coloring it.	They will color in one grid square centimeter.
Pose the problem: How do we cut the paper to make a meter tape from this centmeter grid? How many strips should we cut? How wide can we make it?	Markings on the paper.	They will tell their ideas. They will cut the paper accurately. They will tape it together.

Measurement

INPUT	MODEL	CHECK FOR UNDERSTANDING
I will tell how to number centimeters.	My tape	They will end up with 100 cm.
I will tell how to number decimeters. Ten centimeters equal one decimeter.	My tape	They will end up with 10 decimeters.
Each child should now have a tape one meter long. The end of the tape should read 1 meter, ten cm, and 100 dm. Now we are ready to find our heights in metric measure.	Show poster	Children who have this will raise their hands.

Guided practice

Model how each will measure the other, mark the height with the person's name and tell him or her, "Your height in metric measure is (give several ways in which it could be read)."

Independent

Children will form pairs to measure their heights. Children will record their metric heights on the back of their tapes.

The lesson that could follow this would be to use these metric tapes to compile a Metric Length Page for each child. Children will work in pairs to make specific requested measurements.

Closure

(When time runs out) Roll up your tapes and secure them with paper clips. We will make more measurements with them tomorrow!

Assessment

Observation: Observe how students use their own meter tapes in measuring height. Then ask them for a couple of different ways of reading the height e.g. one meter and three decimeters; thirteen decimeters; 130 centimeters.

Stringing 'Em Along

by Melissa K. Kassis

Using standard units of measure
Converting nonstandard units to standard units

The children have been introduced to the centimeter unit and know how to read centimeter units on a meterstick.

Materials

- 5 metersticks
- 5 pieces of string the length of a meterstick
- Recording charts as shown in lesson

Objective

Using a string and meter stick, the student will measure four body parts in centimeter units and then use those body parts to measure three different objects in the classroom.

Introduction

Start by informing the children that they are going to be learning more about the centimeter unit. Review the centimeter unit on the meterstick. Let them know that every day they carry around reminders of how long different things are in centimeter units and they don't even know it. Tell them that they are going to be measuring different part of their bodies in centimeter units so that if they ever want to know how many centimeters something is, they can easily use a part of their body to figure it out.

Procedure

At this point, break them up into pre-designated groups of three, telling members of each group where to situate themselves in the room. Then, go around to each group and appoint one person as the "recorder" by handing that child a piece of paper which has the following information on it:

Name	Right Arm (cm)	Right Leg (cm)	Right Foot (cm)	Right Hand (cm)

Tell the children that the person who was given the piece of paper is in charge of writing down all the measuring information they are going to gather. That person's title is "recorder." Tell the recorders that they are responsible for writing down the name of each person in their group as well as all the information required on the sheet of paper.

Tell the children that the first person on the sheet of paper listed is to be measured first by the other two group members. Explain that the person being measured is to stand quietly without moving while they are being measured. Tell them the measuring order is on the sheet. First measure the right arm, then the right leg, etc. Then when they are done measuring the first person, do the second person the same way and then the third. Explain to them the measuring procedure. Tell them that both measurers are to measure the other person together. Only one of the measurers needs to take the string up to the meterstick to measure its length and they are to determine who is to do that. At this point, visually show them exactly what they were doing and how to measure by using one group as an example and going through the procedure with the first person on their sheet. This way, they can see how to measure the person with the string, how to take the string up to the meterstick, how to determine the length of the string, and how to record that information. Have approximately five metersticks situated around the room so there would be only two groups at a time working with a meterstick. Also, assign each group to the meterstick they are to use. Then ask for questions.

Exploring

At this point, the groups should be busy doing their measuring and recording. Be available to clarify procedures and to monitor group activity.

Summarizing

Ask the children what they found out about the length of their body parts. Could they use that information to determine the length of something else in centimeters? Tell them that this is what they will be doing next time. Discuss experience working in a group.

Follow up lesson

Have the children make their own individualized chart of the information they gathered in the previous lesson. The chart could look like this:

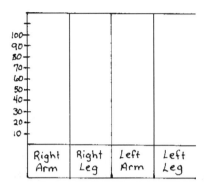

Have them bar graph their body part information

Object Measured	Body Part		Length of Object Measured
	Name	Length	
1.			
2.			
3.			

Then, give each student a sheet of paper with the following information on it and instruct them on how to measure three different objects in the classroom using different body parts and how to calculate the standard length of those objects by adding up how many times they had to use the body part to get the length of the object.

Have the children show their addition on the back of the sheet of paper for each object they measured as this will indicate comprehension of the task given them.

Assessment

Journal writing: Explain how you can use your own hand to measure the length of an object in standard units such as centimeters.

Round Things

by Susan Macaluso

Concept of π

Children will measure and compare the diameter and circumference of several round objects using string and will record the ratio in a table.

Materials

Mini-trampoline, boxes, four round objects per group e.g. cup, can, plastic container, plate, scissors, pencils, charts, string.

Introducing

Every day we see many things that are round, either outside, in the classroom, or at home. Can you name some round things? (wait) Thanks! Today we are going to discover the relationship between two parts of a round object. These parts are the diameter, which is the distance across the object, and the circumference, which is the distance around the object. (Use a mini-trampoline in the center of the floor to demonstrate.) Do you have any questions? In a few minutes we are going to measure the diameters and circumferences of some objects to see how they are related.

Now I will explain how you will do these measurements. You will work in groups of four with each group having a box of four round objects and some string in it. You will also have one recording chart per group. First, take the string and measure the diameter of one object and cut it off. Then take another piece of string and measure the circumference of the same object and cut it off (Model correct and incorrect ways to measure diameter and circumference.) Finally, lay the two pieces of string on your desk and see how many times the short string fits alongside the long string. (Demonstrate these steps using the mini-trampoline.) Measure only one round object at a time and then record your findings in your charts. The chart works this way: the mea-

Objects	Short String	Long String
	I	
	I	
	I	
	I	

surement for the short string is already filled in as "1"; all you need to do is write down the object and the number of times it took for the short string to fit beside the long string. (Record an example on the table) Any questions?

As for working in groups, there are several jobs that will need to be done: someone to measure, someone to cut the strings, someone to write in the chart, and someone to act as the delegate for the group to tell the class what you discovered. I will let the members of each group decide who does what. We also need some rules: you may talk among yourselves quietly, you may ask me a question only when every member has the same question, and if your group finishes early, you may write on your sheet what you learned about circles from measuring diameter and circumference. Are we all clear on the jobs and the rules?

I will divide the class into groups according to their assigned seats. Then I'll give each group a box containing four round objects (e.g. a plastic lid, a can, a tupperware container and a plate) a chart, a pencil, a pair of scissors and ample string.

Exploring

The students will be working in their groups. I will go from one group to another to observe, listen, and check charts. Check to see that the children are measuring the diameters and circumferences correctly.

Summary

Ask each group to relate its findings through its delegate. Discuss the fact that three times the diameter equals the circumference. Ask the students if this is always true or if this is true only in the US, if it's true for large round objects or only these small ones. (Review mini-trampoline exercise, doing it again with help from students.)

Discuss working in groups. How did you like it? How did a group decide who did which job? How did you feel about that job? What things are necessary for a group to work well?

Note: The equation $C=3d$ (approximately) or $C=\pi d$ may be introduced at this time or in a subsequent lesson. Older students can be introduced to the concept of π by measuring with standard units (e.g. inches or centimeters).

Assessment

Journal writing: What did you find out when you used the short string to measure the long string? Do you think this will always happen when you measure circles no matter where in the universe you are and no matter how big the circle is?

Measurement

Bigfoot
by Pat Marshall

Measuring area using standard square units and estimating area using nonstandard units

Objective

Given a piece of centimeter square graph paper, students will measure the approximate area of their hands and feet. They will also estimate the area of other irregular shapes, e.g. Bigfoot's foot, using their own hand as a nonstandard measuring unit.

Materials

Two centiment square grid transparencies (appendix) for the teacher and two grids and a pencil for each student.

Introduction

Tape a "tracing" of Big Foot's foot on the front board (approximately 41 centimeters long and 15 centimeters wide or 16 inches x 6 inches according to the World Book Encyclopedia). Discuss whose foot it is supposed to be. Ask what they have heard about Bigfoot, Sasquatch, and the Abominable Snowman and if they think any of them are real. Read what the encyclopedia says about Bigfoot.

Review what a square is and call attention to the fact that the tracing on the board is pretty big! We could find out how big by finding out how many squares it covers. Color in a square inside the foot to show them the size of a square centimeter. "This square is one centimeter by one centimeter." Ask for estimates of the size of the entire foot in square centimeters and record the estimates on the board. (You could also do the same using a square inch as the unit of measure.)

Point out that it's easy to use squares to find the area of a square or rectangle using a grid but it is also possible to find the approximate area of a circle or any other shape by placing it on a grid paper, tracing around it, and then counting up all the squares that are at least halfway inside the shape. Model tracing the foot on the board by taping two adjacent clear grids over the foot and counting every square that is at least half in the foot. Do this for several rows in the foot and discuss if a square looks half inside the foot. Write "Total number of squares: _____" and its name, Bigfoot, inside the foot.

Their task, in pre-selected groups of two, will be to find the "area" or "total number of squares" for Bigfoot's foot. They will do this by measuring their own hands (with fingers closed) or feet or both in square centimeters. They will trace around the area to be measured then color in and count all the squares that are more than half inside the tracing. (Tell students whether you want them to leave their shoes on.) The student with the longest first name will begin by measuring and recording the other partner's hand, then they will switch. "With your partner, figure out a way to make an estimate of the size of Bigfoot's foot."

Have the paper monitors pass out two sheets of grid paper to each student. Check for understanding of the activity by having two different people explain what they will be doing during exploring. Remind them of your signal to stop talking (e.g., a bell) and of any rules of behavior you want to emphasize during exploring (e.g., cooperation, courtesy, 12-inch voices). "You'll have 15 minutes to complete this activity. When you and your partner each have your grid papers, you may begin."

Exploring

Students in pairs will trace each others' hand and foot. Then they will find the areas of each. Circulate to see how they are conducting their groupwork.

Write an extension activity on the board for those who finish early. "When you finish, look around the classroom and list the things you see with the same approximate area as your hand, for example, an eraser."

Signal the end of exploring (e.g., bell) when everyone is done or time is up.

Summarizing

"What did you get for the area of Bigfoot's foot? Explain how you and your partner arrived at your estimate." Record their new estimates. "Did anyone get a different answer? Did anyone do it a different way?" Count the actual number of squares inside the foot and record the answer. "Which of our earlier estimates came the closest?"

Measurement

Another extended activity

Metric scavenger hunt. The class must search through the room and find objects with the approximate area of:

> 2 sq. cm.
> 10 sq. cm.
> 25 sq. cm.
> 50 sq. cm
> 200 sq. cm.

Assessment

Journal entry: Natasha wanted to find a shortcut for finding the area of an irregular shape like a swimming pool so she thought of a plan. She would measure the distance around the swimming pool by laying a long rope completely around the edge of the pool and cutting it off where it met up with the beginning of the rope. Then she would take that length of rope to the cement and make a square out of it. She thought that if she marked off the number of square meters inside the square, she would then know how many square meters big the swimming pool was. State whether you think she has a good idea. Then, using your grid paper, string and pencil, draw a swimming pool and prove your answer.

Slicing Rectangles to Find Area

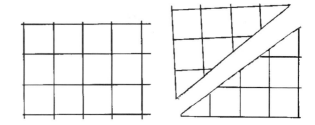

by Jennifer Olds, Jennifer Haskins and Dani Doiron

Area of a triangle

The students will discover that when a rectangle is sliced along its diagonal, it forms two triangles whose areas are the same. This discovery will help students to use and understand the standard formulas for finding the area of triangles.

Materials

Two pieces of centimeter graph paper, geodot paper or geoboards, colored pens, and for the teacher rectangles and triangles of different sizes already cut out and a transparency of the graph paper.

Transitional activities

The students are already able find the area of squares and rectangles with some understanding from previous lessons.

Classroom environment

This lesson should be taught in groups of four children. They will be quietly listening during the instructor's explanation and then they will be using ten-inch voices during their group work.

Anticipatory set

(Show them two different sized rectangles). Raising a quiet hand, who thinks that rectangle has a greater area? (Show them a rectangle) Who thinks that this rectangle has a greater area?

Measurement

Purpose

Today you will be learning about slicing rectangles in order to determine the area.

Procedure

1. The teacher will demonstrate for the students how to find the area of a region by counting the number of square units in the region. Show rectangle, parallelogram shape on transparency.

2. The teacher will then show the students two right triangles in the rectangle and ask them to find the area of each triangle. Show them that the triangle is half of a rectangle. Show them that if you count the square units of the rectangle and 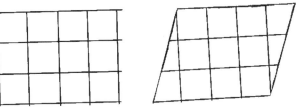 divide by 2 (or cut the number in half) you have the area of the triangle. Label the base and height of the triangles. Have the students try this with several different sizes of rectangles e.g.. 6 square units, 12 square units, 18 square units.

3. The teacher then gives two problems for the students to solve:
a. Given an area, create a rectangle to match it.
b. Given an area, create a triangle to match it.

4. Students share their results of one of the problems. One representative from each group of four goes to the overhead and shows the cut out paper rectangle or triangle on the grid transparency and explains how she was able to get the area. For example, "I wanted a triangle with an area of 6 so I made a rectangle with area 12 (3x4) and I sliced it diagonally."

Assessment

Observation: Using geoboards or geodot paper have student make a rectangle and find its area. Have student record its base, height and area. Then have student slice the rectangle in half diagonally to make a triangle and guess what its area will be. Then have student count the square units inside the triangle to test her guess.

CHAPTER 4
Probability and Statistics

Probability Detectives   Impossible, certain and probable events   K-5
Graphing My Favorite . . .   A generic plan for graphing   K-4
Flavor Favorites   Collecting and graphing data   K-5
Our Pets   Making a picture graph   K-2
Graphing Eye Color   Collecting and graphing data   K-4
Coin Tossing Activity   Recording a random event   1-6
The Jolly Rancher Draw   Predicting from the results of a sample   3-8
Seeing Red and Feeling Blue   Non-random events   3-8
Beans! Beans! Beans!   Graphing data   3-8
What's in the Bag?   Predicting outcomes   3-8
Probability   Graphing the distribution of a random event   5-8
Handedness   Graphing a phenomenon   5-8
Fair and Unfair Spinners   Likely outcomes on spinners   5-8
Math For Health   Statistics: Mean   5-8
Beat the Teacher   Sample space   6-8
Fun With M&Ms   Statistics: Frequencies and mean frequencies   5-8
Roll of the Dice   Probabilities and odds   6-8

Probability and Statistics

Probability Detectives

by Marion Steed

Given 3 bags each with specified unifix cubes, children discover the meaning of impossibility, certainty and probability.

Materials

For Each Pair of Children

- Paper bag
- Red unifix cube
- Blue unifix cube
- Scissors
- Glue stick
- Individual chart
- Red and blue crayons

For Teacher

- 3 bags
- Red and blue unifix cubes
- Red and blue crayons
- Class Chart (See below)

Anticipatory set

What do you think I will pull out of this bag?

Introduction

Today we are going to learn to be detectives and figure out what color cube I will pull out of the bag using the clue: what's inside. If I put 10 red unifix cubes in this bag, what color can I take out? How many times will I pull out a red cube? (Always) (Model - shake and count, 1-2) Do it and record on Class Chart. Introduce the word "certain." Everyone say the word together.

Repeat using blue cubes. How many times will I pull out a red? (0) Record on Class Chart showing all blue. Introduce the word "impossible."

Introduce the bag with one red and one blue cube. How many times will I pull out a red? Do it twice and chart it on Class Chart. Do you think you'll get more blue or red? Which side of your chart will fill up first? Form pairs of children, one to grab cube and one to record the color on the chart. Distribute to each pair a paper bag containing a blue cube, a red cube and an individual chart. Model the jobs.

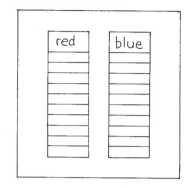

Exploring

Have the children choose a cube from the red and blue bag and chart it. Continue doing this until either the red or the blue side of the chart is filled. (Emphasize that they are to stop as soon as one side is filled so that you get an accurate count). Have the students raise their hands when they are done. Have those who are finished cut out the colored parts of their charts and glue on the class chart.

Extension

If they finish doing all this they can try again on a new chart (race car) and see if the same color would win again.

Summarizing

Ask how many got more blue? Red? Look at the class chart and notice that together we almost had the same number red as blue. Why? If we changed the number of red and blue cubes would it make it different? We can be

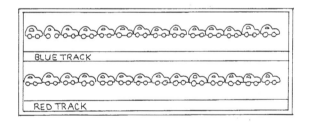

detectives another day and find out. Have the children tell by looking at the chart from which bag it would be a certainty to choose a red cube. Post the word CERTAIN. Repeat with impossible and probability.

Assessment

Close-ended question: Into which of these three bags would you reach if you wanted to be certain to get a red cube?

Probability and Statistics

Graphing My Favorite...

by Susie Beiersdorfer

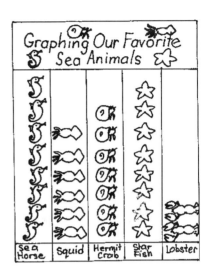

Note: The lesson taught and presented below is "Graphing My Favorite Sea Animal" to coincide with the curriculum the students were learning. Think of this as a generic lesson with many "favorite" applications (ice cream, numbers, colors, days of the week, etc.)

The students will construct a class graph by organizing their choices of favorite sea animals and will also produce their own graphs. They will use the class graph and their own graphs to review the concepts of most (favorite), least, how many more than or less than and other interpretations of the organized data.

Materials

- Large piece of butcher paper for graph (blue paper with picture and name of each sea animal in columns)
- Small cards (name tags for class graph)
- Paste in margarine tub
- Worksheet for each child (Attached is a generic worksheet for graphing in primary grades. It holds up to five columns for categories. Fill in the name of the topic you are graphing.
- Crayons

Anticipatory set

Now for several weeks you've been learning about sea animals. You've learned about starfish, sea horses, squid, hermit crabs and lobsters. Today we're going to make a graph showing which sea animal is the most favorite...least favorite. We will see how many people who choose each sea animal.

Introducing

Here's our class graph...Let's read these sentences at the top: "We can read graphs. We have studied sea animals. My favorite animal is:"

O.K. Is your favorite animal a . . . ? I want you to all choose your favorite sea animal and I'm going to hand out a small piece of paper for you to write your name on. I'll call on people, a few at a time, to come up and put a little paste on the back of the name tag and then to paste the name tag on our class graph under their favorite sea animal. While you're sitting there quietly you can be thinking about which animal you think will be the class favorite. After you've pasted up your choice I'll give you your own graph. We'll work on this together but while you're waiting for everyone to paste up their choice, you can draw a picture of the sea animals above their names on the graph.

Guided practice

The teacher calls each child individually by name, asking the question "Which is your favorite sea animal?" assisting them in placing their name papers on the graph sheet, and giving them an individual graph to fill out when they are seated. Also monitors those still seated. "Which one do you think will be the favorite for the class?"

Summarizing

O.K. Now that everyone's chosen their favorite sea animal . . . Which animal is the most favorite? Did you think the sea horse would be the most favorite? (Thumbs up or down) Which animal is the least favorite?

Now we are going to mark down on our own graphs the number of choices under each animal. First let me show you how to mark a choice and an easier way to show 5 choices. (Model tallying on chalkboard ||||) Now this is an easy way to read groups of how many? (Five)

Independent work

Each child copies the class graph using tally marks.

Summarizing

Now let's look at the graphs we made . . . Does everyone have their own graphs made? An easy way to check if you have the right number of marks is to count the marks and see if it's the same as the number of students in this classroom today which is "28". This class graph with your names on it shows the same thing as your graphs with the marks. What is that?. . .(The number of children who like each sea animal.)

Probability and Statistics

Now let's use this information to answer some questions. First let's look at the questions on your graph sheet. Which sea animal is the least favorite? How many people chose it? (Have them raise their fingers, great way to check for understanding) Which sea animal is the most favorite? How many people chose the sea horse? (20 chose this one, Can we show it with our fingers? This was a fun question!)

O.K. now let's answer some more questions and you can use either the class graph or your own graph:

- How many more (most favorite) than (least favorite) are there?
- How many more (sea horses) than (lobsters) were chosen?
- How many less (squid) than (lobsters) were chosen?
- Are there any sea animals that were chosen the same amount?

Let's talk about what things we've learned from making the class graph and our own graphs...Can you tell just by looking at your graphs what the most and least favorite sea animals are? How can you tell? How many is (   )?...

Assessment

Close-ended: *Was your favorite _____ (e.g. sea animal) more, less, or the same as the class' favorite _____ ?*

Open ended: *What else did you find out about what our class likes?* (Later: *Anything else?*)

Source: Jacobson, Marilyn. "Graphing in the Primary Grades: Our Pets." *Arithmetic Teacher* 26 (February 1979): 25-26.

We Can Read Graphs

We Have Studied _____

My Favorite _____ Is _____

Which _____ Is The Least Favorite?

Probability and Statistics

Flavor Favorites

by Janet Hooper

Graphing data

After interviewing classmates the students will tally the class' favorite flavor of ice cream and graph the total using picture format.

Materials

- A transparancy listing all the children in the class (make from class list)
- A Flavor Favorites sheet for each child (enclosed)
- A sheet of 8 colored self sticking circle for each flavor
- A transparency of favorite ice cream flavors for graphing

Introduction (anticipatory set)

Ice cream is one of my most favorite foods. Most people like ice cream and have a favorite flavor. We are going to find out which flavor of ice cream is the favorite in our class. We have 4 flavors to choose from and you decide which one you like best. Once you decide, don't change your mind! (Have large paper cones and a circle of white paper for vanilla, a brown one for chocolate, red for strawberry and a purple paper circle for blackberry. Put the circles on the cones to demonstrate).

Use monitors to hand out sheet with name. Holding up the name sheet explain: you will use this list to mark yours and each person's favorite flavor. If a person you ask likes vanilla put a "V" by their name, if they like strawberry put an "S" by their name, a "C" if they like chocolate or a "B" if they choose blackberry. The code for the flavors are above the names if you forget. (demonstrate)

Check for understanding by asking, "What are we trying to find? When someone tells you their favorite flavor what are you going to do? Do you have any questions?"

"Before we start asking each other, please put your name at the top of the page and make a guess about the class' favorite flavor by filling in the sentence. I believe most people in this room like _____. Model with two children how to poll each other. Copy the names of everyone in our class onto your FLAVOR FAVOR-

ITES sheets." (Children copy names onto their FLAVOR FAVORITES sheet.)

Exploring

Children ask each person in the room what their favorite flavor is. "You will have about 10 minutes." Each child records the flavor (e.g. "c" for chocolate)

Be sure to establish a signal (e.g., a bell or flashing lights) to indicate that exploring is over and it is time to get to their seats. Practice it before they begin.

Summarizing

Turn on the overhead projector showing the transparency FAVORITE ICE CREAM FLAVORS. "Let's see what is the class' favorite flavor. From your sheet count how many liked vanilla best. (Everyone should get the same number.) Do the same for each flavor.

Look at your graph. What kind of ice cream is liked by most in our class? Does your guess match what we found out? Which kind was liked the least? Did any two flavors receive the same number of circles? How many more circles are there for the favorite flavor than for ___? What could we use this information for?

Graphs are a way to formulate information. You have done a good job and stayed on task (if they have). You may keep your sheet with everyone's name. Find out the favorite flavors in your family tonight.

If you would like each child to have the experience of graphing his or her own data, this lesson may be completed the following day using the general idea desccribed below, summarizing the same as above. You will need to run off a graph sheet for each child.

Introduction

Give them instructions for filling out the graph. Demonstrate what they will do with their graph sheet and self sticking circles—hold up graph sheet.

Each colored circle will represent a flavor choice. Count how many "v"s for vanilla and stick that many white colored circles in the column above the vanilla and then do the same for strawberry, chocolate and blackberry.

Probability and Statistics

Check for understanding

Ask, "How will you decide the number of white circles for vanilla? How will you decide how many brown circles to put above chocolate?. . etc. (Hand out graph sheets and circles) Put your name at the top of the graph sheet.

Exploring

Put colored circles on your graph. (about 10 or 15 minutes)

Summarizing

(Similar to previous day)

Self sticking colored circles can be purchased at Wishing Well and PayLess stores.

Assessment

Close-ended: What was your group's favorite flavor of ice cream? List the four flavors from the most favorite to the least. In what ways was this list the same as what we found for the whole class? Different?

Open-ended: Do you think we would get similar results if we went to a different class in our school? Why?

Source: *Arithmetic Teacher*, April 1982, page 12.

Flavor Favorites

I think most people in the room like _____

Vanilla = V Strawberry = S
Chocolate = C Blackberry = B

Names *

_____ _____
_____ _____
_____ _____
_____ _____
_____ _____
_____ _____
_____ _____
_____ _____
_____ _____
_____ _____
_____ _____
_____ _____
_____ _____
_____ _____

* Make a ★ next to your own name.

Our Pets

by Cheryl A. Kreuzer

The students and will collect, organize and display information on their favorite pets and create a tally sheet and a bar graph for interpretation (class discussion).

Anticipatory set

Yesterday we made cut-outs of our pets. Today we are going to use them for our math lesson. We are going to do statistics on our pets. Does anyone know what statistics are?

Purpose/objective

We are going to collect information in the form of a picture and paste it on a graph. This one is called a bar graph. By using the picture, we can see information more clearly and make comparisons.

INPUT	MODEL	CHECK FOR UNDERSTANDING
Collect information from students - what kinds of pets do they have?	Write list on board in unorganized manner.	Students give information.
Ask students questions about list. Example: What pets do students have more of, dogs or cats?	Point to list on board.	Students try to guess answers.
Explain tally sheet.	Draw tally sheet.	
Explain a bar graph - ask questions.	Draw a bar graph (bulletin board size for pet cut-outs).	Students answer questions.
Directions for making student bar graph, tally sheet and paste-up of pet cut-outs for class bar graph.	Show examples.	Ask questions - students respond.

Guided practice

Students make copy of tally sheet and bar graph from blackboard.

Independent practice

Students paste pet cut-outs on class bar graph.

Summary

Ask students questions about the results. Example: For which pets do we have the same number of owners? How many students do not own birds? Which is more, the number of fish and bird owners or the number of dog owners? How many more?

Materials

Colored construction paper, scissors, graph paper, paste, crayons, rulers, butcher paper, a black marker, lined paper, and hand-made animal patterns.

NOTE: Students make pet cut-outs the previous afternoon in preparation for this lesson. This can be done as an art project. Collect the cut-outs and hand them back just before the lesson. Also, when posting the turtle or fish, make sure they are the same height as the other animals in the graph.

Assessment

Close-ended: Which pet is the most common in our class?

Open-ended: Do you think there is any relationship between the kind of house a person lives in (duplex, apartment, etc.) and the kind of pet they have?

Source: *The Arithmetic Teacher*, February 1979.

Probability and Statistics

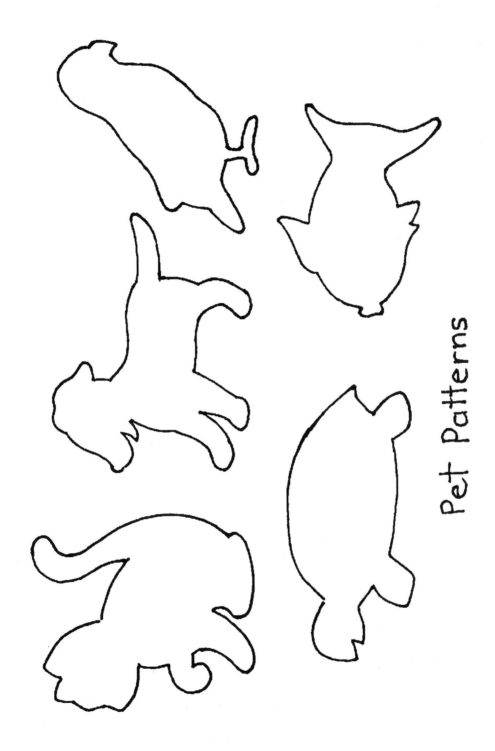

Pet Patterns

Graphing Eye Color

by Kathy A. Sindel

Children will represent the color of their eyes with a cut out and colored eye. The class will create two graphs together; a pictorial and a concrete graph. Then, the class will respond with the correct answers in group response to the questions asked about the four parts of the eye.

A Pictorial Graph

Materials

For each student
- Scissors
- Paste and paste stick
- Pencil and crayons
- Eye dittoed on white paper (4"x2")

For teacher
- Glue stick
- Two graphing charts (see example)
- Eye chart for visual
- Camera and film and flash
- a green, brown, blue, & black colored paper (8 1/2 x 11)

Anticipatory set

Ask the students for words with the word "eye" in it (eye glasses, bull's eye). Ask the students what color their eyes are.

Objective/purpose

Today, we are going to predict what color eyes are the most common in our class.

Probability and Statistics

INPUT	MODEL	CHECK FOR UNDERSTANDING
(What will you say?)	(What will you show?)	(How students will show they know.)
Does anyone know of any words that have the word "eye" in them?	Write their words on the chalkboard.	Students raise have the hands to answer.
Does everyone have the same color of eyes? What colors do you see? Does anyone know what the middle of our eye is called? (pupil) Is it the same color as the rest of the eye? What are the other parts of our eye called? (iris, eyeball, eyelash, eyelid, eyebrow...)	Show visual eye chart and point to the parts. 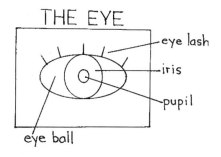	Have a group response of the eyes when you point to the certain parts.
Now, think of what color of eyes are the most in our class. I need a student that thinks we have mostly green eyes in our class to sit here and hold onto this green card. Everyone that thinks there is mostly green sit behind this child in a straight line. (Next brown, blue, and black.) (Teacher stands up on a chair to get a good picture of the whole class.) SMILE! (Take two pictures)		
Now, point to the pupil on your eye. Point to the iris. Look at your neighbor to see if they are pointing at the same part as you.	Pass out the dittoed eyes. Point to pupil on visual.	Students point to part.
If you are not sure what color your eye is, go look in the mirror one at a time. You are to color your pupil black and the iris the color of your eye's iris. Then cut out the eye on the		

INPUT	MODEL	CHECK FOR UNDERSTANDING
outside of the eyeball. Then, come up and paste your eye in the correct column on this chart. When you are finished, come and sit on the rug. (Teacher circulates and checks coloring, cutting, and graphing.) Now, our graph is complete. Will you show me on your fingers how many blue eyes are on the graph. That's correct, it's this many. How many more black eyes are there than brown? Which color has the most? The least? (Smallest) What colors have the same number? Etc. . .	Show concrete graph. OUR EYES Green | Brown | Blue | Black A Concrete Graph Hold up your fingers with the correct number.	Students hold up their fingers showing the number. Students raise their hands to answer.
Can anyone remember what the middle of our eye is called? Raise your hand if you know. (pupil) What is the colored part of our eye called? Tell your neighbor quietly in his or her ear. Now, you can go home and look at everyone else's eyes in your family and see if they are the same color as your eyes. Look at your pets and see what color their eyes are.	Point to middle of eye on visual.	Students raise their hand if they know.

Practice

Guided: The teacher passes out the dittoed eyes. Students point to the parts of the eye that the teacher says. CFU - look at your neighbors to see if they are pointing to the same part as you.

Independent: Students get their materials, look in mirror, color parts of dittoed eye, cut it out, paste it on graph.

Extra

If your class has done a lot of graphing, you may want to make a symbolic graph on the chalkboard or on paper to chart the classes' predictions to compare to the picture that you take. This is great for a comparing lesson later on when film is developed.

Assessment

Using class prediction graph (photo) and the eye graph:
(Open-ended) *Look at these two graphs. What do they show about our predictions and our actual eye colors?*

Source: *The Arithmetic Teacher*, February 1986, "Developing Concepts in Probability and Statistics - and Much More, Bruni and Silverman", pp. 34-37.

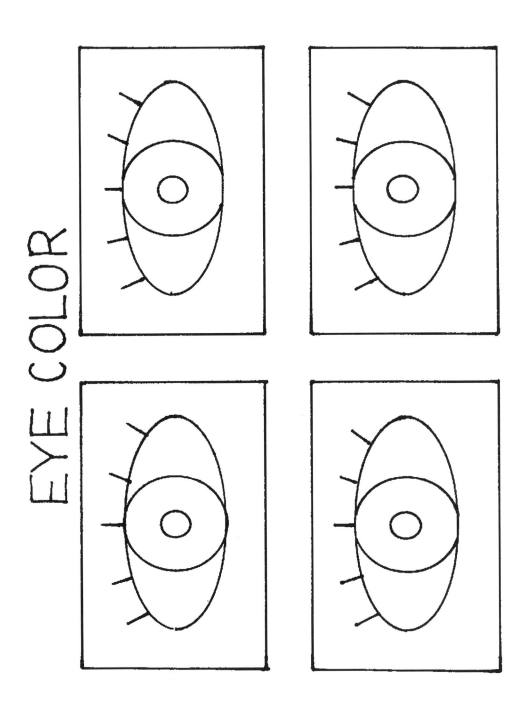

Probability and Statistics

Coin Tossing Activity

by Christy Dalton

Materials

Coins, mats, paper for tall sheet.

Anticipatory set

Ask students if they own or have seen foreign coins. Show some coins in the classroom and talk about them. Then move on to a discussion about American coins. Ask if students can name and identify all the coins in order of monetary value. Talk about what we call the two sides: heads and tails. Discuss instances when people may use a coin toss to make a decision.

Purpose

When a coin is tossed, the result will be a head or tail. We cannot be certain whether we'll toss a head or tail. We can guess, but not be absolutely sure.

Input

We are going to have a total of ten coin tosses.
We will record after each toss whether it was heads or tails on our tally sheet.

Model

Have each student make a guess. Record how many of each on the board. Demonstrate how the coin toss should be done. Shake and drop quietly on a mat. Record the result on a tally sheet on the board. Model how to use tally marks. Each student will make their own tally sheet like the one on the board. Model how to make a rubbing of the head and tails. For younger students, you may want to tape two pennies to a notecard with one of each side facing up. This way the coins will not slide while the kids make their rubbing.

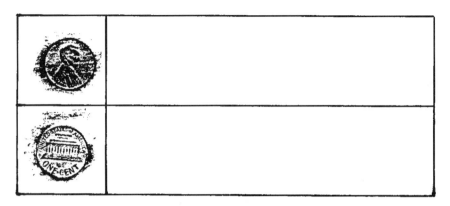

Check for understanding

Give each child a penny and mat.
Check to see that tally sheets are made correctly.
Let students do some practice tosses, making a guess as to the outcome first. Make sure they know which side is heads and which is tails.

Guided practice

Have students toss their coin ten times and record their results. Check to see that they are recording correctly. After everyone has finished, have them report their results by raising fingers to show how many heads and how many tails. Show results using tally marks on the board. Have students add up total number of heads and total number of tails.

Independent practice

Have students continue tossing their coin and recording results. Compare these results to first set of tosses.

Summary

Can we be sure whether we will toss a head or a tail? What do you think will happen if we do the ten tosses again?

Assessment

Journal writing: Have students write the results of their ten tosses each time and the results of the class totals both times. What conclusions do they make about the probability of getting heads?

Probability and Statistics

The Jolly Rancher Draw

by Cathy Evans

Students will demonstrate sampling, random and non-random events. They will make a prediction based on a sample of ten findings.

Materials

3 bags or boxes to hold 8 jolly ranchers each, 18 red jolly ranchers, green jolly ranchers, 3 pads of paper, and 3 pencils.

Anticipatory set

Tell the students that they will be playing a fun new game. "The object of my game will be for each of you to figure out how many of each color of lollipop are in your container." Ask them if they ever wondered how many different colored M&M's were in each package. For example, ask them about the likelihood of getting a blue M&M. After hearing all responses you can proceed to the lesson.

Purpose

Given red and green lollipops, students will figure out what color lollipop they would need to add to their container to make their chances of drawing a green lollipop as good as the chance of drawing a red one.

Input/Model/Check for Understanding

Show everyone the containers. Tell the students that they will be working in pairs and that each pair will be getting a similar container filled with lollipops. Tell them there are 8 lollipops in each container and that they want to guess how many are red and how many are green. Every container has the same number of reds. Tell them that each pair will get a notepad and a pencil. Break the group into pairs. Assign a recorder for the group (one who will record findings) and a reporter (one who will report the findings to the whole group). To check for understanding, ask at least two students what they are going to be doing. You might also ask all the reporters to raise their hands.

Independent practice

1. Give each pair a container (containing 6 red lollipops and 2 green), a pad of paper, and a pencil.
2. The reporter draws one lollipop from the container, has the recorder write down what color was selected and then returns the lollipop to the container. Each group is to repeat this procedure ten times.

Closure

Have the reporters in each group share with the group the number for each color. The teacher records each group's findings. How many red lollipops are in the container? Take guesses and write responses down. Then have students empty out their container. Why did you guess there were more reds than green? What color would you add to make your chances of drawing a red one? At this point ask each pair to write down what color they would add.

Follow-up activities

Let the students draw and record with the same number of reds and greens. From this activity, students will learn how to make predictions based on a sample.

Seeing Red and Feeling Blue

by Julie Hayden Black

Non-random events

The students will be able to:
Collect data
Record data
Interpret data

Materials

For each pair of students
Two red markers or poker chips
Six blue markers or poker chips
Opaque container such as a paper bag
Sheet of paper

Anticipatory set

"We are going to see what color markers (or chips) will show up when we pick them out from the container. See if your guess is correct by watching."

Introduction

"We will be working with a partner and we will be collecting data (what we see). We will each get a turn to be the one who chooses the markers or chips from the container and the one who records on the record sheet. We will record (write down) what we find and later on answer questions about our data (information).

Model the lesson with a student in front of the class first. Place two red markers (or chips) and six blue markers (or chips) in an opaque container. Model the following with the student:

- Tell the class there are some red markers (or chips) and some blue markers (or chips) that cannot be seen in the container.
- Have the student draw one marker. Ask the class "What color is It?" After the students answer, replace the marking pen.

- Have two columns labelled red and blue on the chalkboard or on poster board. Have the student keep a tally of ten choices from the container.
- You can reverse roles and have the teacher pick from the container and keep a tally if the students need more role modeling before they work on their own. Each person will record five results.

Explore

Students work with partners. Each student takes turns at both tasks:
1. Pick markers (or chips) from the container
2. Keep a tally on record sheet

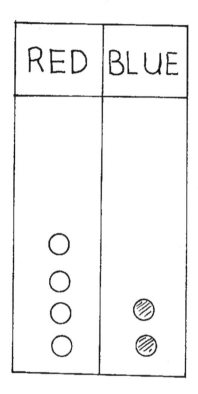

Summarizing

Ask the following questions:

- Did you notice a color that was chosen the most?
- Did you notice a color that was chosen the least?
- Do you think there are more red or blue markers? Why?

After a discussion of how many of each color are in the bag and which color is more, check the guesses by emptying the bag.

Assessment

Journal entry:
Make a game bag with eight chips in it for another class. In this bag make it so that when they pick ten times, most of them will get more reds. Then test out your bag to see if it works. Explain what you did and what you found. Do you think it will work?

Source: "Looking at Facts" by David R. O'Neil and Rosalie Jensen in the *Arithmetic Teacher*

Probability and Statistics

Beans! Beans! Beans!

by Jacqueline A. Barnett

Given 1 ounce (approximately 28 grams) of each of 6 different types of beans, students will estimate the number of each kind of bean and correctly graph that number.

Materials

1 oz. (28 g.) of each of 6 different kinds of beans (pintos or larger work best), student worksheets, wall chart or transparency of worksheet and an overhead projector.

Anticipatory set

Show the students the different kinds of beans and ask who likes certain foods made with beans, such as chili, burritos, etc.

Purpose

"We're going to practice our estimating (good guessing) and graphing. A graph helps us to organize a lot of information so that it is easier to use."

INPUT	MODEL	CHECK FOR UNDERSTANDING
"This is a **pinto** bean. Record that name on your sheet."	Show container #1, record name on chart.	Check that students recorded name.
Ask students to estimate the number of beans in container #1. Explain that estimates are good guesses.	Show size of bean.	Students verbally give their estimates.
"Record your guess on your worksheet."	Record several student guesses on chart.	Students write estimates in correct place. It may help to have students write estimates in pen or crayon.

INPUT	MODEL	CHECK FOR UNDERSTANDING
Give actual number of beans in container and record on worksheet.	Record on chart.	Students fill in actual number in correct column.
Question: "Was your estimate close, within 20 of the actual number?"*		
	Show graphing on chart.	Students correctly graph the actual number.

*(If time is short, you may break the lesson here. Do all of the estimating on one day, and the graphing on the next.)

Guided practice

During the above steps, teacher circulates to check students' work, and answer any questions. If necessary, repeat above steps with beans in container #2.

Independent practice/Assessment

Record on the chart the names of the other 5 types of beans and display the containers. Have students fill in names then estimates for each one. When all students are finished, give the actual numbers in each container. Then have students complete the graph. If students are having trouble completing the graphs on their own, it helps to finish it as a class. To assess, have students make a graph of a different distribution of beans.

Summarizing

Question the students about the graph, in each of these three areas. 1. Use questions to focus students' attention on the graph. "Which beans did we have the most/least of?" 2. Interpret the data on the graph through comparisons. "How many more/less _____ beans did we have than _____ beans?" 3. Question for application of information from the graph. "Would you always expect about this many _____ beans in 1 oz?", or "Does size help you to estimate weight of the beans?"

Source: Adapted from a probability and statistics lesson in the *AIMS Project*

Probability and Statistics

_____
Name

I wonder how many beans I have.

Beans!
BEANS!
BEANS!

Type of Bean	Estimate	Actual
1.		
2.		
3.		
4.		
5.		
6.		

Look at all my beans!!

Graph the number of beans

What's in the Bag?

by Tim Cady

When posed a familiar event, children will correctly determine if a particular outcome is certain, probable, or impossible.

Materials

15 paper lunch size bags, 15 blue poker chips, 30 red poker chips, 30 red-blue charts.

Anticipatory set

Ask if anyone has ever been struck by lightning and explain that it is something that doesn't happen very often. Ask if anyone has ever fallen off their bicycle and explain that the chances of that happening are much greater than being hit by lightening. Tell students that they will now have the opportunity to guess about things that will happen more often than others.

Introduction

Hold a pretend raffle. Only one student bought a ticket for the raffle. Who do you think will win? Model the raffle. Have one student come up and draw the only ticket (Let a red chip stand for the ticket) out of a bag. Next, hold a second raffle. This time two students bought tickets. Who do you think will win? Follow the same steps only this time there is one red and one blue in the bag. Whoever draws the red wins. Explain why the chances of winning this time were slimmer. Next, put two red chips in a bag and one blue. Ask which one will probably get drawn. Draw and record it on the board. Do this two or three times and explain that this is what they will be doing in pairs.

Exploring

Pass out materials and give instructions. Students are to work in pairs. First, one student will hold the bag while the other draws chips and records outcomes on the chart provided.

Then, it's the other student's turn. Each student will do ten draws. Teacher monitors activity and gives assistance as needed.

Probability and Statistics

Summarize

On the board, graph which color won for each pair of students. Discuss why the outcome was the way it was and what might happen if they used all different kinds of combinations.

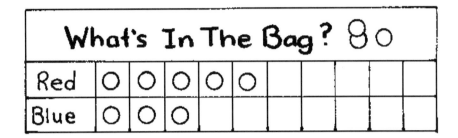

Assessment

Have student draw several lines like the one below with the 0 showing the probability of something impossible happening, the 1 showing the probability of something certain happen and the middle a 50% probability. Have him mark an "X" at the point that shows the likelihood of drawing a blue chip from 10 blue chips if:

1. There are 10 blue chips.
2. There are 0 blue chips.
3. There are 5 blue chips.
4. There are 9 blue chips.
5. There are 3 blue chips.

Probability

by Brandi Zarzana

Given a bag of two cubes - one red, one blue - students will see that the probability of drawing a red (or blue) cube is about 1 out of 2, or 50%.

Materials

For each pair of students:
small brown paper bag
one red cube
one blue cube
2 strips of ten squares
red crayon

Anticipatory set

Pose this situation to the class: Suppose you went to a carnival and saw a game that cost 50 cents to play. The object of the game is to pick a red cube out of a box. If you pick a red cube, you win a dollar. There's a catch - you will not be able to see into the box. Ask the students if they would want to play. Would it be easy to win or hard to win? What would make it easier? Harder? Is there anything they would want to know about the contents of the box?

Introducing

Today the class is going to try to guess which color cube will be picked from a bag that has two cubes in it. One which is red and one is blue. Using one bag, put a red cube and a blue cube in the bag. Have the children watch you put them in the bag.

Have the children guess which cube will be picked from the bag without looking. Whoever guesses that a red cube will be drawn, have them use thumbs up. For those of you who predict a blue cube will be drawn, use thumbs down. Have the students take turns shaking the contents of the bag and taking a cube from the bag, replacing it after each turn. After each turn, ask students if they want to change their predictions.

Probability and Statistics

Talk about how good their guesses are. Are they usually right? How often? Which color is drawn most often? How often will a specific color be drawn in ten trials?

The students will break into pairs ad each pair will be given one red cube, one blue, a handout, one brown paper bag, and a red crayon. Have one student in each pair raise their hand, assign the first jobs to those students. Say, "Your job is to shake the bag, and pick a cube from the bag without looking." Model the job. Check for understanding Next, assign the next job to the other partner. Say, "Your job is to record the result and replace the cube."

Exploring

Model the recording. Say, "The squares will be used to keep track of how often a red cube is drawn. Each square represents one turn. If the cube your partner draws is red, color the square red. If the cube your partner draws is blue, put an X in the square. Check for understanding.

Have each pair of students do the experiment and keep records. Then have the students switch jobs so that they each get to pick the cubes and be recorder.

Summarizing

Tape strips on the board according to the number of reds drawn. Or just make a chart on the board like this:

Talk about the recorded results. Did anyone get red every time? Did anyone never get red? Which result came up most often? 4 of 10, 5 of 10, 6 of 10? How many times did red come up? Do the colors take turns? If one color came up two times in a row, is the next cube likely to be the other color? How likely is it, or what is the probability that red will come up in ten trials?

Extensions

Try this experiment again. Note the cluster of strips around 5 of 10 again. Put more blocks in the bag and perform experiments making the strips cluster in other places. "Can we put reds and blues in the bag so that our graph shows more strips around 2 out of 10?" Develop graphs showing the outcomes of several different experiments.

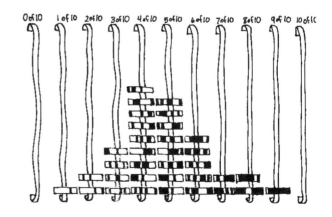

Assessment

Observation: Have some extra recording strips from the experiment, a paper and pencil, and a coin available for the student to use. Ask them to show you how tossing a coin ten times is similar to the cubes-in-the bag game we just played.

Source: Bruni, James V. and Silverman, Helene J., "Developing Concepts in Probability and Statistics - and Much More", *Arithmetic Teacher*, February, 1986.

Probability and Statistics

Handedness

by Patricia Marshall

The class will experience gathering both factual and opinion data and graphing the data. In groups of four they will make a prediction, using their own sample of four as the basis for the prediction.

Materials

A pen or pencil and a piece of blank paper for each group of four

To begin

Enlist the help of two students ahead of time, one left-hander and one right-hander. "Would you sign in, please." Have the class observe, looking for a difference (handedness) as each writes his signature.

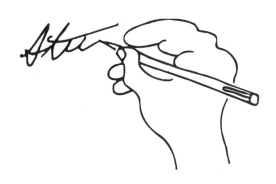

Introducing

"We've been talking about what makes each person special and unique. Handedness is another trait that makes us unique because with this hand we do most things. An interesting fact about the human brain is that it is organized so that the left half of the brain controls the right half of the body and the right half of the brain controls the left half of the body. So left-handers are sometimes referred to as being 'right-brained.'"

"How many students in our class do you think are left-handed?" Record every different opinion and then have the class predict as you record: "How many think there is only one leftie in the class?" Three lefties? etc.

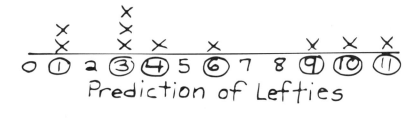

"You're going to get in groups of four to see a sample of the class (your group) sign their names. Based on your sample, you will answer these questions (post these for a reference):

1. What percent were lefties? 0% 25% 50% 75% 100%
2. Based on your sample and any other relevant information brought out by the group, what percent of the whole class are lefties?
3. What facts have you heard about lefties?"

Jobs: Notary gathers the four signatures
Moderator makes sure all group members get to speak on each question.
Recorder takes notes on the answers to the three questions.
Reporter reports to the class.
The group members should come to consensus. (Briefly model the jobs.)

Exploring

Groups are assigned to a section of the room where they will work. If any group has a problem in deciding who gets to do what job, speak to them about the importance of being able to agree. They will need to learn this to be happy and successful in life. Listening, expressing one's view calmly and compromising make groups agreeable. Since we will work in groups regularly, there will be opportunities for "jobs" in future activities.

When the groups have finished signal the end of exploring e.g. turn the light off briefly.

Summarizing

"A science book said that about 10% of all people are left-handed which in our class of 30 would come to about three students. Let's hear what each group thinks about this number." (Record the responses to questions one and two on graphs:

Probability and Statistics

"Let's see our real data. Raise your left hand if you signed your name with your left hand." (Note the number on the board and compare it to the other data.)
"If we surveyed another class, what percent of lefties do you think we would find?"

Discuss the answers to question three. Note that some people are ambidextrous e.g. they write with one hand but pitch ball with the other or they can do some things equally well with either hand. Certain professions seem to have more than 10% lefties e.g. architects.

Follow up

Survey another class and analyze the data.

Assessment

Journal entry: What did you learn about using a sample for predicting percent of lefties?

Fair and Unfair Spinners

by Maria Lamirande

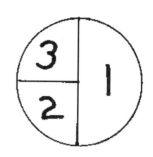

The students will tell orally which is the winning spinner and why, based on the results of their games.

Materials

One Spinner Sheet and three score sheets for each team. Make each spinner using a paper clip, a pencil and a spinner face from the enclosed master. Directions are shown below. Scissors are needed to cut out the four spinner faces.

Anticipatory set

Ask the children questions about their experiences with chance. "Have you ever played a game using spinners? How about dice? Do you believe in luck or do you believe more in a kind of system?" Tell children about the "game" we are going to be playing in teams of 4, 2 pairs per team. Do not tell them about the unfair spinners. They must discover this on their own.

Purpose/objective

This lesson is designed to introduce students to the world of probability and statistics and to introduce new vocabulary pertaining to the subject. Our society deals with statistics and probability everyday from the stock market to weather prediction. The information we get in these ways helps us to make important decisions with some degree of confidence. Key words: fair, unfair, recorded results, statistics, probability, outcome, odds.

INPUT	MODEL	CHECK FOR UNDERSTANDING
Each team should have one spinner sheet, four score sheets, one pair of scissors, four pencils, and four paper clips. Give each team a name.	Hold up the items as you name them.	Have them hold up the items as you name them. Have them hold up their spinner sheets when you call their team name.

Probability and Statistics

INPUT

You will work in pairs. Both of you will spin your spinner at the same time. Do this three times checking your results each time to see who has the higher number. Whoever gets the higher number most often in three spins wins that game. You will play ten games.

After each game, record the results on the score sheet by writing the number of the winning spinner in the box.

At the bottom of each score sheet, write the name of your team and which spinner you had. You will find the number of the spinner above it. Cut out the four spinners and write your team name and spinner numbers on the score sheet now.

Explain how each pair in the team will play 10 games taking turns recording results. A total of 10 games must be played by each pair. When both pairs in a team finish ten games, the winning spinners will play a ten-game play-off. When they are finished, the whole group of 4 will discuss

MODEL

Show how to spin the spinner.

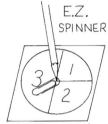

Tape a Score Sheet to front board. Show on Score Sheet where you mark the number of the winning spinner.

Point to bottom of sheet where you have a sample name written. Show where to look above the spinner for its number.

CHECK FOR UNDERSTANDING

Have two or three students from the class show you how to spin the spinner

"If my spinner won, what will I need to write? So and so, tell me how to record that spinner 2 won the third game." Student demonstrates.

Ask some students where and what to write on their score sheets. Have them do this before starting the games.

Ask for questions. Ask some kids how many games they will play. Ask what they will do when they are finished.

INPUT	MODEL	CHECK FOR UNDERSTANDING

the results. They may play a few more games if they have to wait for the rest of the class to finish. Tell them that each team will present their results and they will be expected to come up with some theory about why the results came out the way they did.

Guided practice

Have 2 students in a team spin three times and tell the teacher who won that game. Teacher records spinner number on Score Sheet on board. Remind them that the third Score Sheet is for the play-off.

Independent practice

The game is the independent practice.

Summary

After all the teams have finished their games and have had some time to discuss the results of the recorded information among themselves, bring their attention back to you. Which spinner won in your team? Why? If you played spinners 1 and 4 against each other again, are you pretty sure which would win? (4) Why? (There is a higher probability of getting a 3.) What about spinners 1 and 3? (Outcome is uncertain. It's a fair game.)

Assessment

Predict which spinner will win more. Predict how many of ten games it will win. Then play and record the results of ten games and describe and explain what your results were and why you think you got these results.

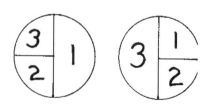

Probability and Statistics

Game Number	Winning Spinner
1	
2	
3	
4	
5	
6	
7	
8	
9	
10	

Score Sheet
Be sure to write down the team name and the spinner numbers before beginning the games.

Record each win with the spinner number. Make sure you record the results as soon as you finish each game.

Circle the spinner pattern that looks like your spinner.

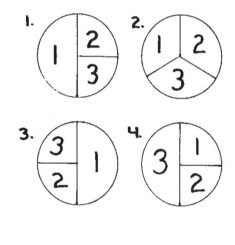

Team Name _____

Spinner Numbers _____

Math Plans

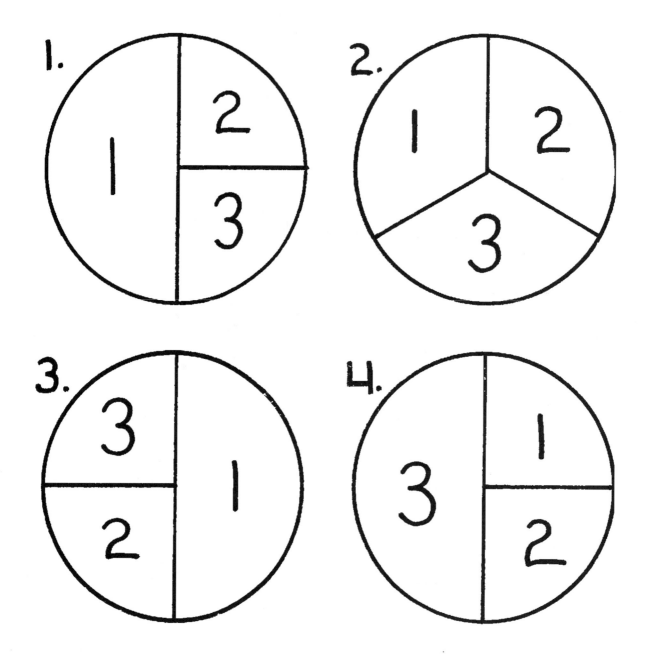

Math for Health

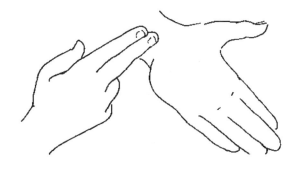

by Anna Brandt Mayfield

Statistics - Mean
Graphing

Students will read a graph, average numbers together, and will graph their own data concerning their heart rate.

Materials

A clock with a second hand or digital stop watch, graph paper for each student, and one large heart rate graph made of poster board.

Anticipatory set

Instruct all students to stand up and stretch. Then ask students to run in place for fifteen seconds. Ask students: "What happens to our bodies when we exercise? What do you think has happened to your heart rate? How can we make our heart rate slow down or quicken?" Instruct students to find their own pulse (either on their neck or wrist).
Circulate the room and assist those who are having difficulty locating their pulse. After students are seated show them your graphed heart rate. Graph these points: 70, 120, 170, 100, 90, 70 and explain how and why graphs are used.

Introducing

Give each student graph paper and have them label the graph appropriately as seen on the model. Ask students what is meant by the "average". (I found that most fifth graders already know how to average numbers together). If they don't, however, explain and illustrate with four numbers. Explain that each person will graph their own heart rate and then we will average them together and put them on the large graph to display in the room. Explain to students that they will be jogging in place for 30 sec., 60 sec., resting for 30 sec., 60 sec., and 120 seconds. A pulse will be taken at those particular time intervals. Have students predict what they think will happen to their heart rate and why. Model jogging, stopping to take pulse for six seconds, multiplying by ten and recording on graph.

Exploring

Have each student find and record their pulse before jogging and then at each interval afterward. Carefully explain that to get an accurate count, they will be counting their pulse for six seconds and then that number will be multiplied by ten.

Summarizing

Ask students what happened the longer that they exercised. What about the longer they waited after jogging? Students should conclude that the longer we wait, the more our heart rate comes down, and that the more we exercise the higher our heart rate will go. "Now let's see how your heart rate compares with the class average." Have two students record everyone's heart rate for the 0 seconds and 60 second jog readings. Let the class find the two averages.

Assessment

Show four towers of unifix cubes - 6,10,4 and 4. Ask, *What is the average height of these towers?"* Then ask the student to show how they would "prove" what the average actually is. (See if they "equalize" the towers, symbolically find the sum of the 6,10,4,and 4 and divide the sum by 4, or verbally reason why "6" makes sense as the average or something else.)

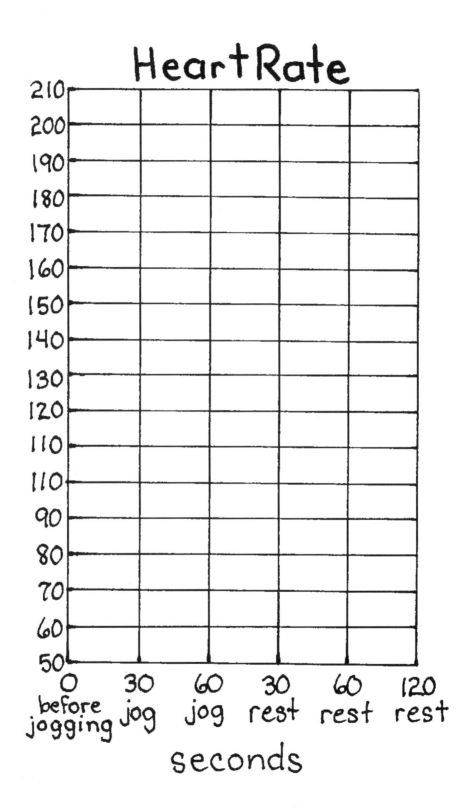

Beat the Teacher

by Julie Richards

Non-random events

Students will realize that the sums of 5,6,7,8, and 9 have a high probability of occurrence when two dice are thrown.

Materials

10 marbles (two different colors, seven of one color, two of another), a small bag, two dice, table of sums shown.

Transitional activities

This is an introduction to probability.

Classroom environment

This is a whole class project so the noise level will be fairly high.

INPUT	MODEL	CHECK FOR UNDERSTANDING
"Many games are based on what is called chance or probability. Probability is the study of making predictions about events and determining how likely events are to happen. After information is gathered, predictions can be made."	"Raise a quiet hand and tell me what probability is."	Pick a quiet hand for the answer.
"We are now going to play mystery marbles. This bag contains ten marbles. We are going to gather information about the color of these marbles. When we have		Have students pick and then return 1 marble from the bag, then

Probability and Statistics

INPUT	MODEL	CHECK FOR UNDERSTANDING
some information we will make some predictions about the marbles."		record the color.
"What are some predictions we can make with this information?"		Pick quiet hands.
	Open the bag and show them the colors.	
"We drew one color more often than the other because there was a better chance to choose it. There is a "higher probability" of choosing those marbles. "Understand?"		Thumbs up/down.
"Now we are going to play a game. You against me."		
"When two dice are thrown, the numbers on the tops of both dice can be added together to get what is called the sum."	Throw dice. Read numbers aloud. "What is the sum of these two numbers?"	Choral response.
"The object of the game is to win by scoring the most points. If we roll the dice, and the sum comes up 5,6,7,8,9 I get one point. If the sum comes up 2,3,4,10,11,12 you score one point."		
"Everyone will roll the dice. You must roll softly, so the dice does not fly across the room. If you throw the dice improperly you will lose your turn"	Show how to properly throw dice. Throw some practice rolls. "Who would of gotten the point?" Point to the winner.	

Practice

Play the game. Make sure to get in at least **thirty rolls**. The teacher will win.

Closure

"Who won the game? Why did I win the game? Why do you think 5,6,7,8, and 9 come up the most often?" Show and discuss the table of sums. Have students shade in all the sums of 5,6,7,8, and 9. Compare the shaded regions to the non-shaded regions. "Who had more information before the game, you or I? Did I win by chance? If you played this game again what sums would you pick? Play 'Beat the Teacher' at home."

Assessment

Journal writing
When we threw the two dice, what sums came up the most often? (5,6,7,8,9)
Why do you think these numbers came up more often than the sums 2,3,4,10,11,12? (There are more ways to get a sum of 5, for example, than 2.)
If we used two 0-5 dice instead of the 1-6 dice, what are all the sums we could possibly get and which sums would occur most often? (0,1,2,3,4,5,6,7,8,9,10,) (3,4,5,6,7)

Source: "Probability in the Intermediate Grades", *Arithmetic Teacher*, Feb., 1979.

Probability and Statistics

Fun with M & Ms

by Andrea Hurlbut

Students will gather real data, find frequencies and organize the data in tables.

Materials

Enough individual packs of M&Ms for each student to have one pack, a copy of chart for each student, one chart transparency, overhead projector, scratch paper.

Anticipatory set

Ask class which colors of M&Ms are favorites and which they think are most common. Tell class that we will determine which color is most common through the activity today.

Introduction

Explain that we will count the number of each color in the packages and determine the average frequency. Explain each column. Review frequency and average. Ask if someone can explain how to find the frequency and average. Model the process for column A. Explain that they will work individually on column A and in groups for columns B and C and that you will lead them as a class for columns D and E. Review rules for group work and announce groups. Be sure each group chooses a reporter. Instruct the students not to eat any of the M&Ms because it will make the results inaccurate. They can eat the M&Ms at the end of the lesson. Pass out M&Ms.

Color of item	a Frequency in your own package	b Frequency totals for your group	c Average frequency your group	d Frequency for totals for the class	e Average frequency for the class
Red					
Orange					
Yellow					
Green					
Tan					
Brown					

Exploring

Tell students to begin working. Set a time limit for finishing the work. Walk around the room to observe and assist students.

Summarizing

When students are finished, ask the reporter from each group to share their group's results from columns B and C with class and record on overhead chart. Discuss the conclusions and do columns D and E as a class. Ask students why they think these 2 colors are most common. Are the common colors their favorite colors?

Assessment

In groups of four, students poll their classmates to collect data on this question: *How many people live at your house/apartment?* Arrange the data in a table.

Source: *Arithmetic Teacher*, 1977. "M&Ms Candy: A Statistical Approach" by Donald Hyatt.

Roll of the Dice

by Kelly R. Timpson

Students will predict the probability and odds of all events involving two dice, graph the results of 50 rolls of the dice, and compare results with predictions.

Materials

A coin for the teacher and enough dice for each student to have one. Pencils and paper (graph paper) if desired.

Anticipatory set

Show the students a coin and ask them what the chances are for a head or tail to show on the flip of the coin. Discuss probability (1 to 2, 1:2) and the odds (1 to 1,1:1) of the possible results, showing symbolically how these are represented.

Have the students predict the outcome of 20 tosses of the coin. Then perform the experiment. Have the students keep a tally of the tosses as you model them on the overhead projector. Compare to probability predictions.

Coin Toss

Heads   Tails

Introduction

Show the students a die and ask them the number of possible outcomes when the die is rolled. Have the students determine the probability of rolling each number. The six-sided die is helpful in further explanation of odds. Give each student a die and have them find out how many rolls it takes to get a one (1). This is a good time to point out that probability predicts an outcome but that it is still a random event.

Now pair off the students and have them roll one die 30 times. Have them choose one to be the recorder and one to roll the die. The recorder will tally the results of each roll. Discuss outcomes.

Exploring

Have each pair of students determine the number of possible outcomes of rolling two dice and the different combinations that will produce any one result. For example, there is only one way to get 2: (1+1) but there are 6 ways to get 7. Have them roll the dice fifty times and record the results of each roll. Then model putting the results into graph form on the overhead. Have them put their results in a graph.

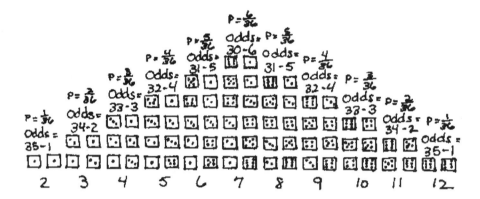

Summarize

Combine the results of all the pairs on the overhead and ask the students to estimate the odds and probability for the experiment, then compare to the actual figures (see above chart). Discuss odds in relation to games of chance and gambling.

Note: Ring a bell to signal the start and end of each activity.

Assessment

Writing: Have students write a response to this situation that includes a drawing, chart, or graph of some kind.

Your best friend says, *"It's just pure chance when you toss two dice. No one can say what number will come up."* In what ways do you agree? In what ways do you disagree?

CHAPTER 5
Patterns and Functions

Show You Know   Reproducing a pattern in a different form   K-3
Stamping Out Patterns   Making patterns using rubber stamps   K-2
Staircase to the Clouds   Complex patterns   2-8
Leftovers   Odd and even numbers   2-4
The Magic Box   Function rules   4-8
Cubes and Ways   Functions   4-8
Looking for Patterns   Functions 4-8
Patterns in Multiplication Facts   Star patterns on a circle   3-8
Powers   A paper tearing activity   5-8
Can You See It In A Name?   Grid patterns   3-8
Patterns and Functions   Paper folding   5-8
Side-Stacked Squares   Patterns in perimeter   6-8
Chips and Dips   Looking for a pattern to solve a problem   5-8
Handshake Problem   Looing for a pattern to solve a problem   5-8
Basic Functions   Organizing input and output data   6-8
Function Machines   Making a function machine from a cereal box   2-8

Show You Know

by Suzanne Lilliedoll

Students will reproduce, extend, describe and translate a pattern to a different form, e.g. concrete to auditory patterns.

An AB Pattern

Materials

Sets of unifix cubes containing at least four different colors. A chart depicting a clap snap pattern with a red cube drawn under the clap and a blue cube drawn under the snap (or any color combination desired).

Anticipatory set

I will begin clapping a pattern and invite children to join in when they think they know the pattern. (snap clap snap clap)

Purpose

Today we are going to practice clapping and snapping some patterns. We will also be working in groups of two to show some patterns using the unifix cubes.

Input

Place the children in groups of two, each twosome sharing a set of cubes. Explain how they will watch me snap/clap a pattern. When they think they know the pattern, they will work with their partners to show the pattern using cubes.

Modeling

Show the chart and explain it. Snap fingers and clap hands. Point to a row of alternating blue/red colored cubes below your hands. Say, "Snap, clap, snap, clap, "as you point. Explain what the snap/clap pattern might look like and demonstrate using the cubes as shown above.

Patterns and Functions

Check for understanding

Clap out a simple "AB" pattern and watch to see that the students understand how to build the pattern with cubes.

Guided practice

Clap and snap a series of increasingly difficult patterns, moving from "AB" to "ABC" to "ABCD" patterns. As you clap each pattern, invite the children to clap with you. When it looks like they know the pattern, have them show it using cubes. Then do it in revrese: you make a pattern using the cubes and your student "partner" describes the pattern verbally and claps it. Remind them to take turns, one makes the pattern using cubes and the other guessing the pattern.

Independent practice

In pairs, let children build patterns of their own design with cubes. Ask for verbal explanation of pattern. Have children try clapping out their partner's cube patterns.

Conclusion

Conclude by having children share their patterns. Compare patterns e.g. "Who else had an "ABBA" pattern like this one?" Note that we were able to show our patterns in lots of different ways. Have children name ways. (clapping, snapping, tapping, cubes, verbally.) But all of our different patterns had one thing about them that was alike. Who can guess what it is? (They repeated)

Assessment

See how students can show you a pattern they recognize. Clap a pattern. Start at the "Describe it" level. If the student can do it, move to "Translate it."
If not, move back to "Extend it." If they cannot extend it, move back to "Reproduce it." Remember that an "ab" pattern is easier than an "abb" so within each level you can vary the difficulty by changing the pattern.

Reproduce it:	Clap/snap an "ab ab ab" pattern.
	To student: "Clap this pattern!"
Extend it:	Clap/snap an "abb abb abb" pattern.
	To student: "Clap what comes next." abb abb a...........
Describe it:	Clap/snap an "aab aab aab" pattern.
	To student: "Tell me what this pattern is."
Translate it:	Clap/snap an "abba abba" pattern.
	To student: "Show me with these red and yellow cubes the pattern I'm making."

Stamping Out Patterns

by Patti Bartholomew

Students will explore known patterns such as A, B, A, B, . . to find new ways of expressing them.

Materials

Prestamped pattern cards (make these up ahead of time), ink pads in two or more different colors, pencils, variety of rubber stamps, paper cut into 3x 8 1/2 inch strips, empty egg carton to hold and separate stamps, scratch sheet for clearing stamps of colored ink.

Anticipatory set

Teacher begins with statement such as, "Since we have all been so successful at recognizing A, B, A, B . . .patterns, its time we had fun exploring rubber stamp pattern cards!"

Introduction

Explain to the students that the object of this game is to discover as many ways as possible to express the patterns. This activity will be done as a work station in groups of five or six students. (This reduces the expense of providing materials on large scale and makes for ease of monitoring comprehension.) Before students may begin, rules need to be stated. 1) No stamping on each other, themselves, or desks! 2) Return stamps to egg carton, rubber side down when stamped card is complete. 3) Clear stamps of ink onto scratch sheet when changing colors of ink! Check for comprehension of rules. Demonstrate an A, B, A, B. . .with the stamps. To use rubber stamps correctly takes practice and skill building. Let students practice by creating patterns of their choice. Pass out materials. Limit students to two stamps each so that patterns will resemble those on pattern cards. (5 minutes for practice) Ask students to share their patterns. Probe to see if they can come up with other ways to express patterns besides A, B, . . . If not, give suggestions such as, patterns with shapes, numbers, animals, colors, etc. (See last page for ideas) If students still need guidance, vocalize these new patterns for them. Collect patterns. Distribute premade pattern cards. Explain that students need to copy pattern exactly as they

Patterns and Functions

see it. Give students new paper strips and have them locate necessary stamps to duplicate patterns.

Be sure to explain that it is more important that they understand how the pattern repeats than to have the card as exactly spaced or as neatly done as the model! Check for understanding.

Exploring

Students will locate stamps and begin. Observe to see if any students have difficulty. If so, have another student explain the activity. Continue until all students have completed one or more cards. (This should take about 5 minutes)

Summarizing

Have students share their pattern ideas, one at a time. When student has exhausted ideas, ask others if they can think of another way to verbalize the pattern. If students fail to come up with more than two or three, ask probes again such as, with color, shapes, animals, words, numbers, etc. (10-15 minutes) Have students write their names and the patterning possibilities on back of strips. If they need assistance, write it for them. Ask students if they were surprised at how many different ways of patterning are possible, even in short, simple patterns. They will begin to recognize how important it is to try to view things in a variety of ways.

Extension

Have students explore pattern making with three or four stamps.

Source: Original idea brainstormed by kids' love for rubber stamps!

Assessment

Using her own pattern, have the student describe the pattern in as many ways as she can. If she doesn't use a number pattern, ask if she can describe it using numbers.

SAMPLES

Ice Cream, Joy; Food Word; 1, 3

2, 3; Bear, Gingerbread Man; Toy, Cookie

3, 2; Smiling, Not smiling; Faces, Bunnies

Patterns and Functions

Staircase to the Clouds

by Paulette Johnston

Complex patterns

Students will work in pairs to construct patterned staircases of at least six steps with unifix cubes.

Materials

Unifix cubes, graph paper

Classroom Environment

The students will work in pairs so there will need to be discussion between pairs as well as whole group discussion. The noise level will be medium.

Anticipatory Set

"Today we are all going to be architects and construct staircases that could be tall enough to touch the clouds." You can discuss what architects and construction workers do. You can also discuss the purpose of staircases, especially that they go up, up, up.

INPUT	MODEL	CHECK FOR UNDERSTANDING
Constructing a staircase is easy because it involves a pattern. A pattern is something that repeats itself. The steps repeat so that the staircase will grow higher and higher.		
Let's see if you can guess the pattern I am making with the unifix cubes.	Teacher begins to build one step pattern.	Ask for quiet hands to guess what will the next step be the pattern. Call on two students to do the next two steps.
	Repeat the above step using a different staircase pattern.	

Guided practice

"Super Job! Now it is your turn to build a staircase with your partner." have students work in pairs constructing staircases of at least six steps.

Discussion

Ask students if they could build the stairs higher? How high could they go? How would the staircase look? Talk about the pattern and how patterns help us to know what comes next. With our imaginations, this pattern of stairs could really take us to the clouds.

Independent practice

"Nice job working with your partners. Now let's get started on making individual staircases. Using your graph paper, draw a staircase of at least six steps. See how many different staircases you can make before our time is up."

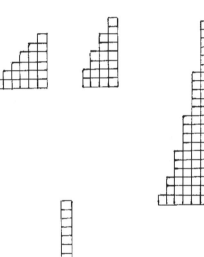

Assessment

Ask students to choose his favorite pattern from the class creations, describe the rule used to make the pattern and then describe the pattern using numbers e. g. (4, 8, 12, . . .)
(2, 2, 4, 4, 6 . . .)

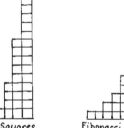

Other Patterns

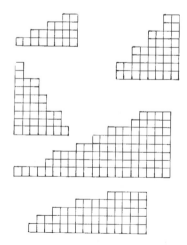

Leftovers

by Elizabeth Anne Brothers

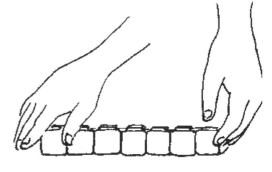

Odd and Even Numbers

After playing the "leftover" game in pairs, the children will describe to the teacher and the other children the pattern they discover on their chart and predict whether or not the other numbers under twenty fit this pattern by circling the "yes" numbers on the bottom of their paper.

Materials

Each pair of students will need 9 unifix cubes, 1 paper with chart, and a pencil. (Use the model shown to make a master).

Anticipatory set

Ask if anyone in the group knows what leftovers are. Get children to realize that leftovers are not necessarily food but anything left over. Tell them that they are going to investigate leftovers in groups of two and let each pair decide between themselves which one wants to be the does/manipulator and which one wants to be the recorder. If children have trouble deciding roles, quickly decide the roles for them. Tell them they each need to be checkers of the other to make sure that the jobs are being done correctly.

Introducing

Have the doers quietly get nine unifix cubes, a pencil, and a chart. When children are in pairs, have the doers line up six cubes and put their hands on each of the cubes on the ends. When you say "grab", they grab the two end cubes and put them in the middle of the table. Then say: "Ready, grab. Ready, grab. Are there any leftovers? Then write "No" after the 6 on your paper." You can model this with unifix cubes and then check for understanding by having them try the procedure with seven cubes. The recorder can be the one to tell the doer when to grab. Make sure they understand that they write "yes" if there are leftovers and "no" if there are not and that they record the answer by the number they <u>began</u> with.

Exploring

Have the children do this same process in pairs for the other numbers on their chart. Ask them to be looking for patterns as they record the results of their investigation. Have them circle all the "Yes" numbers on the number line at the bottom of their paper and challenge them to see if they can predict which other numbers on the number line will be "Yes" numbers also. (This could be an extension) The teacher should monitor each group to make sure they are understanding. Special attention should be paid to the way the children decide whether or not 0 and 1 have leftovers. Simply asking them "If you have nothing can you have anything left over?" usually suffices for 0. If they have problems with 1, make sure they understand that both hands have to have something to grab each time or else it is a leftover.

Summarizing

When the children have had adequate time to finish their charts, have them share their findings with the whole group. What kind of patterns did they see? Which numbers have leftovers? Could anyone predict which other numbers would have leftovers. In the discussion, you could introduce the terms "odd" and "even".

This lesson was designed for first graders. If older children are doing this, groups of four would be effective and the children could make their own charts instead of the teacher providing them.

Assessment

Prediction: Show student a stick of five unifix cubes and ask if there would be any left over. If he says, "Yes," commend him and ask the same for six and eight. If he says, "No," let him test it out himself and then ask him again. If you want to assess vocabulary (odd and even), use these words in the assessment.

Leftovers

0	5
1	6
2	7
3	8
4	9

0 1 2 3 4 5 6 7 8 9
10 11 12 13 14 15 16 17 18 19

The Magic Box

by Pamela Howe and Christie Sonmez

Functions

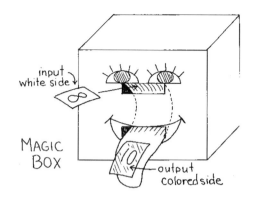

Given a number line template and corresponding numbers from the magic box card, students will correctly write 2 out of 3 addition equations and circle the patterned number in each of these equations.

Materials

- The Magic Box*
- Magic Box Cards*
- Number Line Templates*
  (one for each child)
- Pencils
- Lined Paper

* Directions attached

Anticipatory set

Today we are going to play a game with the "Magic Box". With enthusiasm the box is uncovered and viewed. Explain to students that the box can perform the magic that the teacher whispers to it. The box changes the numbers on a card as it passes through the box and out of its mouth.

Purpose/objective

We are going to learn how to write addition problems using a number line template (show it) and the magic box. Then we are going to look to see how each problem is the same in some way.

INPUT	MODEL	CHECK FOR UNDERSTANDING
Explain the number line template and the circle holes beneath each number.	Show the template. Trace the circle below the number 1.	Have the students trace the circle below the number 1.

Patterns and Functions

INPUT	MODEL	CHECK FOR UNDERSTANDING
Remove the template. How many circles have you drawn?	Show your example.	Check to make sure each student drew one circle.
Explain that the magic box cards have numbers on them. We can check the answers from the Magic Box using our number lines.	Show a card. The number shown is 2. Show that you draw the number of circles shown on the card.	Have the students draw 2 circles.
Describe the magic of the box.	Feed the card with the number 2 to the magic box. Show the colored side of the card. The number is 5.	Ask students what happened to make sure they were watching.
We want to add enough X's to the number 2 until we get to the number 5.	Show how to draw X's in the holes to the number 5.	Students add enough X's next to the circles to get to the number 5.
Explain what is now drawn.	Show your 2 circles and 3 X's.	Have students look at your example and tell you how many circles there are and how many X's. Look at their papers. Make sure all have 2 circles and 3 X's. If not help them get there.
Explain the addition equation.	Point to your equation and say, " 2 and 3 is 5."	Have students repeat chorally," 2 and 3 is 5."
Explain the written equation.	Write the equation and make sure you tell them that the = sign is another way to say "is".	Have students write the equation. $2 + 3 = 5$

Repeat steps 1-8 if students have not grasped the concept. Use different numbers to model the process. The function "rule" in this example is "+3."

Guided practice

As the cards are shown the students draw the correct number of circles and add the correct number of X's to the circles to reach the number on the colored card. The teacher monitors the drawing of the circles and X's and checks the written equation for correctness.

Teacher shows each card and feeds it to the magic box. The students write 3 more equations without difficulty. Teacher checks to make sure at least 2 equations are written without major errors.

Summary/closure

Teacher asks how all of these addition equations are alike? (eg., They all add 3.) The students are asked to circle the number that is the same in each of these problems. Discuss why they circled that particular number. The students assess what they did in each of the problems.

Assessment

Have student make a deck of cards that uses a "rule" she has chosen (e.g. +4) to use in the Magic Box. Assessment can be based on whether all the cards in the student's own deck follow the same rule.

Note

The Magic Box is used in all grades using increasingly difficult "rules" as you move up the grades. Let students make their own deck of cards for the Magic Box and have the other students try to guess their rule. Younger children benefit by seeing that addition and subtraction are opposites, one undoes the other, by using the box in reverse. Third- and fourth-graders will find the same for multiplication and division. Older children will enjoy squaring numbers and using two "rules" or more on each number in their deck of cards.

Source: Baratta-Lorton, Mary. *Mathematics Their Way*. California: Addison-Wesley, 1976. Pp. 248-249.

Directions for the Box, Card and Template

The Magic Box

Made from any size or shape box. Cut out two rectangles (3-1/2" x 3/4") in the front of the box. Cut a "tongue" (3-1/2" x 18") out of red poster board and tape it to the top edge of the top rectangle. Allow the "tongue" to stick out and then tape it (from underneath) to the box just below the bottom rectangle. Quantity needed: one for each group.

Magic Box Cards

Each set of cards should be a different color. They should be plain on one side indicating the "in" side and colored on the other side. The number on the plain side indicates the "in" and faces the child when it is fed into the magic box. The "out" number on the colored side is the number that results when the rule for this particular set of cards is applied to the number on the plain side.
Sets of cards can be made to follow many differerent rules. For this lesson on patterns and functions, patterned examples below demonstrate the rules for the sets, "+3", "+4" and "+5".

Two Cards from a Deck of Cards for F+2=B

Number Line Templates

A number line written on heavy tag board 2" x 11" with holes under each number (cut with a pair of manicuring scissors, a circle cutter, or punched out with a hole punch.) The child draws circles or X's in each appropriate hole. Quantity needed: one for each child.

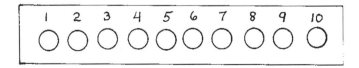

Cubes and Ways

by Debra R. Caldwell

Functions

- Work in groups of four and problem solve.
- Discover a pattern.
- After working with concrete materials (unifix cubes), describe orally and in written form how many ways a given number of cubes can be arranged in two groups.
- Record in a table the number of cubes and the number of ways.
- Use a number sentence to describe relationships of a pattern in two piles.

Skills

Counting, comparing, using a pattern, examining numbers, physical exploration of a pattern and materials (unifix cubes), making and testing predictions and using small muscles to manipulate the cubes.

Materials

- A set of 10 (ten) unifix cubes for each group of four (Have more available off to the side for use with extensions)
- One sheet of paper with the Cubes and Ways table (a sample table is shown in this lesson) and a pencil for each group of four
- Chalkboard and chalk for the teacher or a large piece of paper and a marker
- The Table: How Many Legs are on 8 Chickens? This is used as a warm up activity.

Introduction

The teacher describes what a pattern is. "A pattern is a plan, diagram, or model to be followed in making things." Also the teacher will talk about how patterns are the basis of our number system and how it works; patterns make it possible to predict (guess) what is supposed to happen in mathematics and finding a pattern is another way of solving a problem. As a warm up activity the teacher will do with the children the problem, How Many Legs Are On Eight Chickens? First, make a chart on the board as follows:

Patterns and Functions

<table>
<tr><td colspan="2" align="center">How Many Legs on 8 Chickens?</td></tr>
<tr><td>Number of Chickens</td><td>Number of Legs</td></tr>
<tr><td>1</td><td>2</td></tr>
<tr><td>2</td><td>4</td></tr>
<tr><td>3</td><td>·</td></tr>
<tr><td>4</td><td>·</td></tr>
<tr><td>·</td><td>·</td></tr>
<tr><td>·</td><td>·</td></tr>
</table>

Then ask the children, "How many legs are on one chicken?" Fill in the chart with the number 2 and then proceed to "How many legs are on two chickens?" Fill in the chart with the number 4. Proceed on through number 8 each time filling in the chart on the board. After doing the first two or three ask the children if they can see a pattern in the numbers that will help you complete the chart. (The pattern is: number of chickens x 2 = number of legs). Using the children as models, act out a similar problem: How many legs do eight children have?.

Divide the children into groups of four. Explain the rules for the groups: 1) "You are responsible for your own behavior." 2) "You must be willing to help <u>anyone</u> in your group who asks." 3) "You may not ask the teacher for help <u>unless</u> all four of you have the same question." Assign roles in the group and explain what each person does and that each person will be taking a role during the activity. The roles are as follows: <u>Doer</u> - gets materials needed by the group; <u>Recorder</u> - writes down the results; <u>Arranger</u> - manipulates (moves) the unifix cubes; <u>Checker</u> makes sure all the numbers are being recorded.

Statement of problem

"Suppose you have ten unifix cubes. How many <u>different</u> (emphasize different) ways could you arrange ten cubes into two <u>different</u> piles?"

Before you begin the activity have each group make a prediction (guess) for the 10 cubes. Write the guesses on the board to be talked about during the summary. Make a chart on the board and model the first two or three ("With just one cube there are two ways; with 2 cubes, you can arrange them 3 different ways. . .").

Cubes	Ways
1	2
2	3
3	·
·	·
·	·

With two cubes the question may come up as to whether a different arrangement of colors or location within a "pile" constitutes a different "way" eg., "Is red, blue different from blue, red?" Answer that they are considered the same since in this activity we are interested simply in the <u>number</u> of objects in each pile.

Before beginning, ask someone to restate the problem and make sure it is understood by all the groups.

Exploring

Students assume their assigned roles of their groups of four. The teacher becomes the observer. Allow children to manipulate cubes to discover a pattern. Check (monitor) that all students are staying on task with the cubes (using them as instructed), that all children are staying in their assigned roles, maintaining the group of four rules, and working together to discover a pattern. If needed, model again for those students and groups that may need the problem retaught. Praise the children for their work, attempts and efforts.

Extensions

- "How many ways can you arrange 15 cubes into 2 piles?"
- Allow children to make other patterns with the unifix cubes through free exploration (These will vary, depending on the children and their creativity).
- "How many ways can you arrange 10 cubes into three piles?"

Summary

Bring all smaller groups of four into one larger group. Ask, "What patterns did you discover while working on the problem?", "What did you find out?". Talk about how many ways it took to arrange the 10 cubes. Ask, "Does anyone know a short cut to finding the number of ways? (You add one to the number of cubes to give you the number of ways). "What were some of the things learned about patterns today?" Make known how the children solved the problem by having the children help you fill in the table you began the lesson with on the board. Finally, return to the children's predictions (guesses) made before the activity began and compare to the answer that each group actually obtained. "Now let's see what you learned. Suppose you had 20 cubes, how many ways could you arrange them in two piles?" Ask several of these which will require "counting up" by one. Praise the children and comment on what good learning you observed and what a good job they did in learning about patterns.

Patterns and Functions

Notes

If I were to conduct this activity again I would spend some more time on Groups of Four activities as these children are not used to working in cooperative learning groups. It would be necessary and appropriate to spend a period or more before this activity to work on groups of four rules and introduce other group of four activities such as ones mentioned in Meyer, C. and Sallee, T. (1983), <u>Make It Simpler.</u> They point out that it is important to have practice with groups of four using classroom problem-solving activities, curriculum problem-solving activities, and logic activities. Practice with groups of four activities involves seating the children in groups of four, establishing and processing with the children the rules for group work and introducing the activity needing a partner. I agree with their approach, especially if children aren't used to working in groups of four in cooperative learning groups. Fifteen minutes was needed for exploration when I taught it because some groups needed to be retaught the process of arranging the cubes in two stacks.

Assessment

Grades 2 - 3 : Have student draw a picture to illustrate the problem and write two or three sentences to explain it. Use this writing prompt:
"This picture shows _____. I think the answer is _____ because _____."

Grades 4-8 : Use the cubes and ways example. Say,"If I told you there were 30 cubes , how many ways would there be?(31) How about 100 cubes? (101.) So then, what are you doing to the number of cubes to find the number of ways? (Adding 1.) Suppose I said there were "C" cubes, "C" standing for a number I don't know yet. I think I will say there are C+1 ways for "C" cubes. Write a paragraph stating whether you agree or disagree with me and explain why.

Looking for Patterns

by Gina Trzaska and Kristy Nettleton

Materials

Paper and Pencils
Geoboards and rubber bands

Anticipatory set

- Start by showing the above pattern on the board leaving a couple of blank spaces for the students to fill in the missing designs.
- Ask for volunteers to fill in the blanks for the pattern.
- Ask students how they figured out what went in the blanks
- Ask what number patterns they see.

Introducing

Warm-Up Activity: "Rule in the Machine"

- Drawn an Input/Output table on the board.
- Select a student to be the "machine".
- Explain to the other students that you will whisper a rule into the machine's ear (the student).
- The rule can be addition, subtraction, multiplication or division (e.g., +10).
- Ask the students to give you 5 different numbers between 0-100 to put into the machine. This is called the "input" and should be put under the "Input" part of the table.
- Explain that the five inputs that were given will be used by the machine to get five outputs.
- Have the machine write the answers under the Output part of the chart.
- Have the students guess the rule in the machine. The students may want to work this out on paper or discuss it with a partner.
- Ask for the rule that the machine was given. Take all suggestions from students. List them on the board.
- Check the rules given by the students to see if they work with the input and output.
- Give them a hint that it may be easier to see the rule if the numbers in the table are rearranged so that the Input part of the table is listed in numerical order. Rewrite the table this way.

INPUT	OUTPUT
2	
15	
99	
⋮	⋮

Patterns and Functions

- Have the machine state the correct rule. Ask the student that got the correct rule to say how he found the answer.
- Reinforce the idea of looking at the information in a table and looking for a pattern.
- If time permits you may repeat this process with another machine.

Main Activity: "Handshakes"

# People	# Handshakes
1	
2	
.	
.	
.	

- Start the activity by asking, "If everyone at the table shook hands with everyone else, how many handshakes would there be?"
- Have the students shake hands with each other, starting with 2 people, then 3 people, then 4 people, etc. Until everyone at the table has shaken everyone else's hand (approximately 4 people at each table).
- Model how a group of three would work before you let them go on their own to shake hands.
- Ask the students what would happen if there were only one person at the table?
- Draw a table on the board representing the handshakes

# People	# Handshakes
1	0
2	1
3	.
.	.
.	.

- Have the students continue the chart by figuring out how many handshakes there would be for 4 people, then 5 people at the table.
- Next represent the number of handshakes by using Geoboards. Each marked peg represents a person and each rubber band represents a handshake. Have students experiment using the Geoboards to see the number of handshakes.

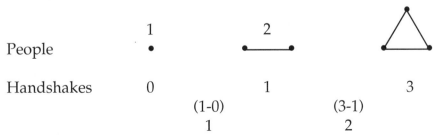

The difference increases by one each time.

Conclusion

Conclude by having the students write their answers on the board and discussing the patterns. One of the patterns you may want to point out is that when you subtract the first number of handshakes from the second number of handshakes and the second from the third and so on you get numbers that increase in order from 1 to 2 to 3, etc... (See example above).

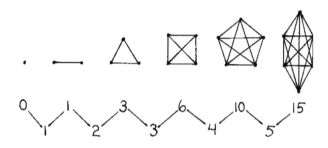

Assessment

Interview: Ask student to explain what she found when her group started to do the handshakes, first with two people, then with three, and so on. At what point did she begin to see a pattern? What pattern did she see? (Description will be based on the actual handshaking or based on the number patterns she saw on the recording sheet.)

Source: *Make it Simpler* by Carol Meyer and Tom Sallee

Patterns in Multiplication Facts

by Theresa J. Olmscheid

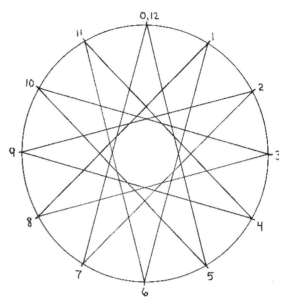

The students will draw star patterns in circles and color grid patterns on a hundreds chart using multiples of 3, 4, 5 and 6.

Materials

- For each pair of students: Hundreds Chart and two grids (See attached samples. I used 1" x 1" squares on 7" x 5 1/2" poster board grid decorated with drawings or stencils.)
- 100 1" x 1" paper squares in two colors, 50 of each color.
- 8 dice.

Anticipatory set

Motivate the students with a discussion of patterns. What are patterns? How can one tell that something is a pattern? Have students relate some patterns they have seen in everyday life. Some examples are egg cartons, linoleum or tile on the floor and a printed piece of fabric. The majority of student responses should be successful because of the variety of patterns in our world.

On the blackboard place three large circles numbered clockwise 0-9, 0-11, and 0-12. (This can be done before the lesson begins. See attached samples of completed circles.) In review of previous lessons, explain that multiplication is a series of identical sets linked in a continuing pattern. Using one of the samples given in this lesson, draw lines on the circles for the multiples of one of the patterns. For example, note the multiples of four on the 0-9 circle. The child starts at zero and counts to four making a dot at four. Then he counts four more, makes a dot at eight and connects the dots. Then he continues counting four and connecting with the previous point until the pattern repeats.

Have two students come up to the board individually and have them draw the lines of the patterns for the circles while the class directs them with line placements for either of the multiples of two or three on the 0-11 circle or either of the multiples of five or seven on the 0-12 circle. Question the students as to what type of patterns they can see. How are the patterns different? The students should notice that patterns are seen in many ways including numbers.

Introduction

Relate that the multiples and patterns that they have just worked with are found in the numbers 1-100. Visually present the Hundred Chart to the class and explain that they will be covering numbers on the chart with squares of colored paper in patterns that are the multiples of two through nine. Demonstrate this, modeling to the entire class the placement of squares for several of the multiples by using the blackboard or an overhead projector. Break the students into groups of two where there is a collector/arranger and a thinker/checker. The collector/arranger will collect a Hundred Chart and a packet of two different colored squares from the front of the room by table order. He will then arrange the colored squares into patterns on the chart. The thinker/checker will think of different patterns to arrange and after they are arranged, he will check to see if they appear to be a verifiable pattern.

Procedure

Have the collectors obtain the Hundred Charts and the packet of colored paper squares per group. The students will cooperate in the arrangement and determination of patterns with the squares to cover the different multiples of two through nine. Let them plot these as high as they are comfortable and let them experiment with their own interpretations. Have them discuss whether they agree that all their attempts are patterns. How are some of the patterns different?

Exploring

Let the students perform their duties within their groups. They should be working cooperatively. Monitor their progress in understanding, success, time and "on-task" behavior. Note any ideas or problems that arise for use in the summary of the lesson.

Extensions

1. Have the collector return the Hundred Chart to the front of the room and have him obtain a die and 5" x 5" grid with the multiples of twos and threes noted on them. There should be one grid per two students and one dice per grid. This game functions along the rules of Bingo. When the die is rolled, this number should be multiplied by two and the product on the grid covered with the student's own

Patterns and Functions

colored paper square. The next round, after both players have rolled once and placed their squares, will have the dice amount multiplied by three. Each round will alternate between two and three. The students get two tries to answer the problem correctly and then they forfeit a turn. Whichever student covers a whole column or row first will be the winner.

2. If the students complete Dice Game #2 several times and if time permits, they may try Dice Game #1 which uses the same rules except they are working with the multiples of four and five.

Summarizing

Bring the students back into one large group and discuss what types of patterns they created and how did they go about deciding that these were the correct patterns for their problems? Did they all get the same patterns? How were they different and what were they thinking about as they constructed their patterns? Were there some groups who still did not understand the lesson? Did everyone see some sort of success in the experience? A correct display and discussion of some of the more confusing patterns can be presented. It should be related that multiplication tables show a definite pattern and some partially coincide with others e.g., 2 and 4.

Suggestions for follow-up lessons

Patterns in Circles

The pattern can be used as worksheets for the students to complete with crayons or markers. Students can be creative by combining and trying their own patterns. They may also work well as examples of manipulatives for working with geometry concepts.

Hundred Chart

The chart can be duplicated onto 8 1/2" x 11" paper so students may color it the different multiples with crayons. It can also be used like a crossword puzzle in which the students color in answers to a series of math questions. When they finish they may see a picture of a happy face, the word "Hi", or a mathematical equation for an operation they are currently working with. The Hundred chart is very versatile and may be used in the study center of the classroom for a free time activity.

The Dice Game

It need not be used solely as an extension. It is a good fun practice of the multiplication tables. It can be expanded to include a progression of larger multiples or a combination of large and small multiple. This is also a good game to have in the classroom study center.

Assessment

Journal Entry: Which star pattern do you like the best? Why? What did you notice about this pattern? What patterns did you notice in the hundreds chart?

Sources:
Cook, Marcy. "IDEAS Section", *Arithmetic Teacher,* February 1989, pp. 31-36.
Tyler, Jenny. *Multiplying and Dividing Workbook,* Usborne Publishing Co., London, England, 1986 pp. 5, 8-9. This book is based on *Multiplying and Dividing* a book by Annabel Thomas and Graham Round.

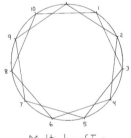
Multiples of Two

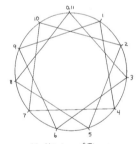
Multiples of Three

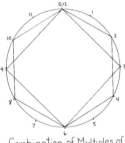
Combination of Multiples of Two and Three

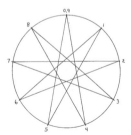
Multiples of Four

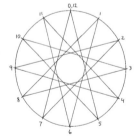
Multiples of Seven

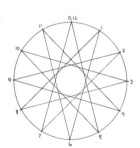
Multiples of Five

Hundred Chart

1	2	3	4	5	6	7	8	9	10
11	12	13	14	15	16	17	18	19	20
21	22	23	24	25	26	27	28	29	30
31	32	33	34	35	36	37	38	39	40
41	42	43	44	45	46	47	48	49	50
51	52	53	54	55	56	57	58	59	60
61	62	63	64	65	66	67	68	69	70
71	72	73	74	75	76	77	78	79	80
81	82	83	84	85	86	87	88	89	90
91	92	93	94	95	96	97	98	99	100

Dice Games 2's & 3's

18	9	8	3	15
4	12	18	9	10
15	8	3	2	6
12	6	16	4	2
10	3	4	18	9

Dice Games 4's & 5's

10	8	20	5	20
25	12	15	24	16
16	4	30	10	8
24	20	5	20	30
15	16	12	25	4

Powers

by Patricia Marshall

Exponents
Functions

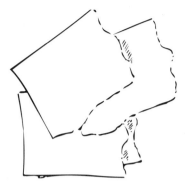

Different ways of tearing up paper yield different functions. Students tear paper and record the results in a chart. They describe the patterns verbally.

Materials

Pencils and paper (for recording)
Paper (for tearing)

Get Ready

"Watch and listen." Tear a piece of paper into three equal strips. Then stack them holding all three in one hand. "How many times did I tear?" (Callouts: "2.") "How many pieces do I have?" (Callouts: "3.") Take a fresh piece of paper and this time tear it into four equal strips. Ask the same questions. Pose this question: "Suppose I wanted to rip this paper into 16 small pieces, how many tears do you think it would take?"

Introducing

The activity: "Many people rip paper into small pieces like this." Take a fresh paper and tear it in half. Stack one piece atop the other and tear in half. Then stack these four pieces of paper and again tear in half. Each time you tear ask the question, "How many pieces now?" Look for a pattern in the numbers that will enable you to predict what numbers will come next.

The recording: "You will be working in groups of four and each group member will get to be a 'tearer.' One of the group members will be the recorder. That person will record the results on a chart showing the number of tearers and the number of pieces resulting each time." Model the recording (see next page).

Patterns and Functions

The jobs: The **getter** gets the paper and is the first to tear.
The **predicter** predicts how many pieces will result and is the next tearer.
The **recorder** records the results on the chart and is the third tearer.
The **counter** counts the pieces each time, does the final tearing (if it is possible!), and throws away the scraps.
If two people in a group want the same job they may resolve the conflict using rock/paper/scissors, best of three gets to choose.

Exploring

The paper is gathered and the predicting, tearing, counting and recording begin. When every group member can describe a pattern in the numbers the group may begin another sheet.

Summarizing

Choose a member from each group to report what the group found. Did anyone find a relationship between the number of tearers and the number of pieces?

Assessment

Interview: You did some paper tearing and some number writing. What did you learn today that makes you feel like a better mathematician?

Patterns and Functions

Can You See it in a Name?

by Cheryl L. Noack

Students begin seeing a relationship among different occurrences having the same underlying pattern. The students will understand that the same pattern can emerge from a variety of settings.

Materials

- Grid paper

Anticipatory set

Who can tell me what a pattern is? Show a calendar page. This is an example of a pattern. What patterns are there?

Look around the room and see if you see some patterns. I see one: the floor tiles. Raise your hand if you can find more. (List on projector.)

Input

Patterns come from a lot of different things. We can even make patterns out of our name.

Modeling

Draw a 4 X 10 matrix on the chalkboard. Write someone's name in the matrix. Color the first letter of the name. See example.

Check for understanding

How could you describe this pattern (checkered, diagonal, etc.)?

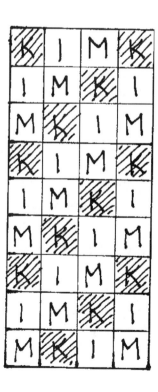

Guided practice

I'm going to have you try this with your name, but first let me assign each person in your group to a job. (Show jobs on projector).

- Supply Analyst - comes up front and gets grid paper for group and makes sure everyone in group has a colored pen.
- Questioner - writes down 4 questions and when your group has finished with the name patterns, leads questioning.
- Recorder - Writes down answers to questions.
- Quality Control - Makes sure everyone is on task, and gets time to share his name pattern with group.

Four questions to discuss with your group

1. What other names have the same patterns? Why?
2. If you put numbers in the squares instead of your name, and colored in the same squares, what numbers are colored in?
3. Predict what numbers would be colored in if you continue the pattern.
4. What if you colored in the last letter of your name. Try it!

Independent practice

Students write their names in grid and color first letters.

Summarize

Discuss Questions:

1. What patterns did you find? Whose names had the same pattern? Why?
2. What happened when you put numbers in the boxes instead of your name?
3. Could you predict what #'s would be colored in?
4. What happened if you colored in the last letter of your name?
5. Would the pattern change if the grid were 3 x 5?

Assessment

Writing: "Did you enjoy making these patterns? Why do you think we did this in math? What do these patterns show?"

Patterns and Functions

by Gary T. Winegar

Recognize number patterns that exist in the real world and record them in table form.

Materials

Two pieces of paper, pencil, chalk and chalkboard.

Anticipatory set

Say: "Have you ever wondered if math can be made easier? I am going to show you one way..."

Objective/purpose

After students receive instruction in performing patterns and functions activity, they will be able to recognize those patterns that exists in numbers and mathematical problems.

Procedure

Begin by drawing a recording chart on the chalkboard for students to replicate onto their piece of paper:

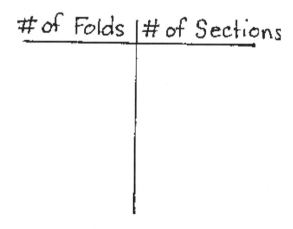

After students have finished recording chart have them pick up another piece of paper and have them fold it in half, modeling for them and then checking for understanding. Have them then unfold the paper and record on their charts: one fold --> two sections. Model this on the chart on the chalkboard. Have the students refold the paper in half and then fold it in half again so they have two folds. Then have them unfold the paper and mark on their charts: two folds --> four sections. Model this on the chalkboard.

# of Folds	# of Sections
1	2
2	4

At this point have the students continue folding the paper, marking on the chart the results after each fold. Have them fold the paper up to six times. Go around the classroom answering any questions and checking for understanding. After students have completed the activity go back to the chart on the chalkboard and ask for the

F # of Folds	S # of Section.
1	2
2	4
3	8
4	16
5	32
6	64

answers to complete the chart. Check for right answers.
Ask the students if they see a pattern existing on the chart. (Answer: For each fold the number of sections double.) Show the pattern again to the students emphasizing

Patterns and Functions

the strategy one could take to find out how many sections there would be if you folded the paper ten, fifteen or even twenty times, without actually folding the paper to find out.

Extension

Depending on the grade level, ask the students for an equation for this pattern, writing them on the chalkboard and examining each one for correctness. $2^F=S$ (This is only for students who have been introduced to exponents. Otherwise, ask them to verbally explain how to get from the number of folds to the number of sections)

Evaluation

Checking for understanding would come during the discussion and answer period after the activity. If further evaluation is necessary have the students continue with the folding of the paper, using it as a manipulative for further understanding.

Assessment

Do this lesson after. Then ask: How is this activity like the paper ripping activity in **Can You Find the Powers?**

Side-stacked Squares

By Mary Charlesworth-Eggers

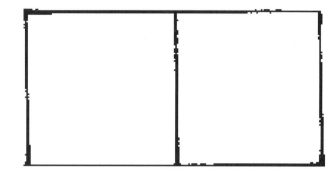

The children will:

- Use visual observations, spatial logic, and will figure out patterns, and
- Will become familiar with the concept of perimeter as they manipulate and experiment with cut out squares, fill out a column graph, discuss their feelings

Materials

- Chalk and Chalkboard
- About 56 (or less depending on how many groups of four you have; there should be 7 squares per group) cutout construction paper squares (1" x 1"), a picture of a square field surrounded by a fence, a picture of a simple house plan.

Anticipatory set

Say to the class, "I want to draw a square on the board. Who can tell me what I should draw to do this?" Draw the square according to the children's directions. Look at the square and say, "So four sides make up the outer edge of a square. Does anyone know another name for the outer edge or boundary of an object? . . .(Wait for a response.) . . . The perimeter of an object is that object's border; therefore, a square's four sides are its perimeter because they show the square's outer edge or border." Show the class other examples. Hold up the picture of the field and point out that a fence outlines a field's perimeter. Hold up the picture of the house. Tell the children, "Pretend that you are sitting on a cloud and looking down at a house with no roof. What outlines this house's perimeter? . . .the walls. Good! You've got a good idea of perimeter."

Introduction

"Now if the 4 sides of a square make up its outer edge, how many sides make up two joined-together squares' perimeter?" Draw the picture on □□ the board and ask the children, "How many sides make up this object's perimeter?" Call on a

Patterns and Functions

child. If he/she answers 6, say "Great!" and count the sides off with the class, then go on to three squares. If the child says 7 or 8, remind the class that the sides inside the shape are not counted. The outer sides are what define the boundary and perimeter. Ask again. When a child answers 6, go on to three squares. Once you feel that the children understand, have them break into groups of four. Model the four jobs that are to be assigned: "doer", "manipulator", "recorder", and "quality control", and then assign these jobs to the children in the groups.

Once the children are in their groups, model a column graph, and have the recorders copy it down for their group. One side is labeled "# of squares", and the other side is labeled "# of sides in the outer edge (perimeter)."

Number of Squares	Number of Sides in Outer Perimeter
1	4
2	6
3	8
⋮	⋮
7	

Have the doers pick up 7 squares for their group. The manipulator must put the squares together so that the other group members can decide on the number of sides and so that the recorder may write down their observations. Have the class redo the 1, 2, and 3 square models by recording the results on the model graph on the board and on the graphs in the groups. Tell the groups to go on and record the results for the 4 through 7 squares, and when they finish, they must raise their hands.

Exploring

Walk from group to group to make sure groups are doing the activity.

Extension exercise

If a group finishes early, have them experiment with rearranging the squares and seeing if this effects the number of sides in the perimeter. For example:

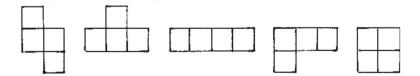

Summarizing

When all the groups have finished, ask the recorder from one group to write their column graph on the board. Ask the oth4r groups if they agree with the first group's findings. Ask the children if they notice any pattern in their findings. (patter: the # of sides increases by 2 each time a square is added) Could they figure out, now, how many sides would make up the perimeter if they had 10 squares joined together even if they don't actually work with 10 cut out squares? (answer 22 sides) If they answer 20, ask them where they think they went wrong. The children quickly realize that they must start counting at 4 and not 2 when they count up to the 10th square.

Extension summary

Ask the group what they found: rearranging the squares does not effect the number

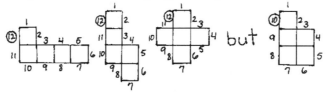

of sides in the perimeter <u>except</u> when they are arranged in a box-like shape. With the box shape you always get two less sides showing than when you arrange it any other way.

Remember

Always be supportive and enthusiastic when presenting the lesson plan. Compliment them especially when they discover the pattern and use it successfully. If the group easily finds that you get 22 sides with 10 squares, you might want to say, "But I got 20" and have them <u>help you</u> discover where you went wrong.

Assessment

Demonstration: If you were designing a hall to display pictures and you wanted to put each picture with a yard distance from the next and you wanted to have the most pictures you could, would you design a square gallery or a long gallery? You will have 50 square yards of space. Prove that you answer is correct.

Patterns and Functions

I. M. Square

by Mary Simpson

Square numbers

The learner will show concretely what a square number looks like using a "flat" or hundred square.

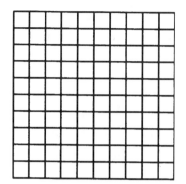

Materials needed

- 1 base ten flat per 2 students
- 2 3 x 5 index cards per 2 students
- 1 Hundreds Chart per 2 students
- 1 crayon per 2 students
- Sum Fun transparency

Introducing

The lesson is introduced by two mental math problems using patterns and non-patterns. More may be added.

Begin with 9 . Multiply it times itself (81). Add the digits (9). Multiply times 2 (18). Add the digits (9). Multiply times 3 (27). Add the digits (9). Multiply times 5 (45). Add the dig8its (9). What is your answer? Work the problem out loud for the whole class. What was the answer every time we added the digits of a multiple of 9? (9). Is there a pattern?

Take the number of letters in the word PATTERN (7). Multiply it times itself (49). Add the digits (13). Multiply times 2 (226). Add 4 (30). Divide by 3 (20). What is your answer? When you multiply 7 times itself and add the digits, did it add up to 7? What about other multiples of 7? (It works for 9, but not for 7.)

Today we're going to explore patterns we find in math. What are some patterns you can think of? (2-4-6-8. . ., 3-6-9. . .).

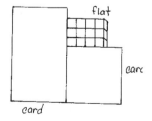

Each group of 2 has one flat (from the base ten blocks) and 2 index cards. Tell students that when you multiply certain numbers together you can model the results by forming a square. Demonstrate by sliding the index cards out showing 2 x 2, 3 x 3, and 4 x 4 squares.

Have the students perform this also. Check for understanding by viewing the results on their desks. Instruct the students to see which of the following could be modelled showing a square.

	3 x 3		6 x 6		10 x 10	9 x 8		4 x 4
3 x 7		8 x 8		5 x 6		7 x 7	5 x 5	

Exploring

Allow 15 minutes for a fast group and 20 minutes for a slower group. Students will record the results on a plain piece of paper. Model result for 2 x 2 on board.
(2 x 2 - yes)

Summary

Which equations could make a square? What do those have in common? (When both factors are the same, it makes a square). Show the students that a simplified way to show 5 x 5 is 5^2. Explain that this is different from 5 x 2. Let students think about it before continuing.

Extension: Sum Fun

Pass out hundreds chart and one color crayon to each pair of students. Instruct students to write both of their names at the top of the pair. They are to work cooperatively to answer the questions and color the answers by trading back and forth. Students are to hand papers in when complete. Show "Sum Fun" transparency on overhead.

Notes

The students really enjoyed the Sum Fun activity. It forced them to view problems in a different way. They aren't word problems exactly, and they aren't standard equations, either. They enjoyed looking for the pattern at the end. It was very easy to correct, also! I let the students know which problems they missed.

Patterns and Functions

Assessment

Demonstration: "Use two cards on the base-ten material (flat) to show every size square possible. How big is each square?" Once the student has stopped showing squares ask, " Could there be other sizes of squares on the flat?"

Some (Sum) Fun

Read each item.
Use a crayon to color the answer on the hundred chart.

1. The number 10 more than 27
2. The number 10 less than 54
3. The odd number between 9 and 13
4. Four cents less than a dollar
5. The number 20 more than 28
6. The value of 9 dimes and 2 pennies
7. Two dozen
8. The value of 2 dimes minus 1 penny
9. The number 1 less than 5 tens
10. The number of sides on 7 nonoverlapping triangles
11. The value of 3 quarters minus 2 cents
12. The sum of 6 tens and half a dozen ones
13. The odd number between 91 and 95
14. The number 1 less than 35
15. Half of 36
16. The number with 6 in the ones (units) place and 8 in the tens place
17. The number 1 less than 4 tens
18. The number that is 1 more than the number of sides on 10 nonoverlapping squares
19. The value of 2 pennies and 1 quarter
20. The number of days in February of a leap year
21. The value that is 1 cent more than that of 6 nickels
22. The difference of 80 minus 4
23. The largest odd digit plus the largest even digit
24. The number with two 3s
25. The number between 44 and 71 made with straight lines only
26. 50 + 20 + 1 + 1
27. The value of 7 nickels

Source: *Arithmetic Teacher*, February, 1989.

Chips and Dips

by Nancy E. Casper

Students will manipulate, identify, and tally the number of different ways 15 chips and 2 dips can be eaten, e.g. 14 guacamole and 1 onion.

Materials

15 chips, 2 containers (bowls), pencils, tally sheets, paper to draw solutions on, package of 2 1/2" x 3" index cards.

Anticipatory set

List on the board four of each student's favorite chips to eat. "Do you ever dip your chip into something tasty before you eat it?"

Purpose/objective

Look for a pattern as a way of solving a problem.

INPUT	MODEL	CHECK FOR UNDERSTANDING
Discuss all the kinds of dips. (Choose two favorite dips and label two bowls with their names, e.g., guacamole and onion.	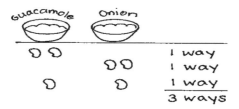	Students receive materials.
"If I had one chip, how many ways could I eat it?"	Dip into guacamole or dip into onion. Two ways.	
Do the same for two and three chips, this time letting children tell you how to arrange chips.	Place both in guacamole, then both in onion, then one in each, etc.	Children direct teacher as to where to dip chips.
This is how your recording sheet will look.	Chips: 1, 2, 3, ... Ways: 2, 3, 4, ...	

Patterns and Functions

INPUT	MODEL	CHECK FOR UNDERSTANDING
Explain students' roles. **Doer:** one who gets material. **Dipper:** one who places chips into the containers. **Recorder:** one who records the results. **Checker:** one who monitors the ways the chips are placed in containers.	Show students what each role requires.	One group demonstrates their roles.

Guided practice

As the teacher models, the students manipulate, tally, and identify ways of dipping chips. The teacher observes the students process of placing the chips into the containers and recording the results.

Independent practice

Students decide which role they will play. Students work cooperatively together and discuss their roles. They also discuss the processes they are using to solve the problem. The students should be able to understand and successfully manipulate and record the number of ways 15 chips can be placed into two containers.

Summary/closure

Lead students in a discussion of their results. "How many ways for 15 chips? What pattern did you find?" Discuss their roles in the group. "Did every group member do his job?"

Extension

Graph the number pairs on graph paper.

Assessment

Interview: "What is the relationship of the number of chips to the number of ways you can dip them?"

See if the student can express this relationship in her own words or, instead, can only repeat the equation or part of the equation e.g. "Add one."

Journal Entry: Have student solve this problem and answer the two questions.

Problem: Suppose you wanted to make double dip ice cream cones for your friends and you bought four flavors of ice cream: vanilla, chocolate, strawberry and peppermint. How many different double-dip ice cream cones could you make? (Hint: A "vanilla/chocolate cone is not "different" from a chocolate/vanilla cone).

Two questions: How is this ice cream cone problem the same as the chips and dip problem? How are these problems different?

Patterns and Functions

Handshake Problem

by Sharon L. Hagen

Problem solving strategies: making a chart
looking for a pattern

Students will record the number of handshakes possible with a given number of people. Students will recognize patterns in the numbers.

Materials

- Pencils (sharpened)
- Lined/blank/graph paper
- Rulers
- Colored pens

Anticipatory set

In 1990 the United States Census Bureau collected information about all the people who live and work in the United States. This census, or information gathering, occurs every ten years. The gathering of all that information is just the first step to learning and discovery about many subjects. There are experts who look at the information and decide what it all means.

Purpose/objective

Today you will gather information in a simple project and determine what it means after you have put it all in some order in diagrams, charts and graphs.

INPUT	MODEL	CHECK FOR UNDERSTANDING
The problem that needs solving is called the Handshake Problem. Ask rhetorically, "If each of you shook everyone else's hand only once, how many handshakes would there be?" The first thing to do is to have a little demonstration. May I have three volunteers?"		Three students stand up to volunteer, after raising quiet hands. One student volunteer is asked to stand alone.

INPUT	MODEL	CHECK FOR UNDERSTANDING
	Teacher points to single student and asks, "How many handshakes can _____ have by him/her self?"	Students respond - many guesses until the rules are clarified: One handshake <u>between</u> each other but one student cannot shake his own hand."

	Teacher puts two students together, has them shake and re-ask the question. Repeat for three students. Record:
	
	# people # handshakes
	   1 0
	   2 1
	   3 3	

Teacher asks if there could be some way to show this problem using a diagram or picture?	Write down student response. If students do not have a workable solution, teacher draws one, e.g., circles with dots representing people and lines representing handshakes.	Student suggest ways to solve the problem.

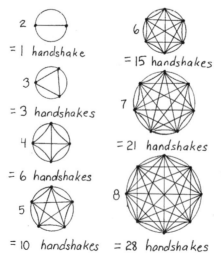

Patterns and Functions

INPUT	MODEL	CHECK FOR UNDERSTANDING
	Teacher shows another way to record information: on a graph. Teacher draws the graph on board/paper. Ask students where the first point will go? Second?	

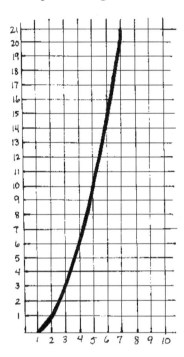

Teacher asks if there are any questions. If not, graph paper is handed out for students to complete work.		Student continue to work together in order to discuss where points should go. Check for accuracy.

Guided practice

Have students draw diagrams or pictures to get them to think about the numbers. Guide them in making the chart for the first three people and handshakes.

Exploring

Occurs as the students work together in groups of two, extending the original problem from eight people to the number of people in the class. As students work together they can check each other for accuracy.

Summary/closure

What did you get for five people? Six? etc. What patterns do you see in the numbers?

Assessment

Journal Entry: Have student solve this problem.

Suppose you came in at the end of a party and there were only two potato chips left but there were three different dips: California onion, guacamole, and bean. If you could dip each chip into only one dip, how many different "flavor pair" choices (e.g. bean dip/ guacamole dip) would you have to choose from? Suppose there were six different dips, how many choices would you have then? Which flavor pair would you choose?

Two questions: "How is this problem different from the Handshake Problem? How is it the same?"

Patterns and Functions

1 ☺ = 0 handshakes
2 ☺☺ = 1 "
3 ☺☺☺ = 2 "
4 ☺☺☺☺ = 3 "
5 ☺☺☺☺☺ = 4 "
6 ☺☺☺☺☺☺ = 5 "
7 ☺☺☺☺☺☺☺ = 6 "
8 ☺☺☺☺☺☺☺☺ = 7 "
                  ―――
                  28 handshakes

☺ = 0 handshakes

☺—☺ = 1 handshakes

△(3 faces) = 3 handshakes

▢(4 faces) = 6 handshakes

(5 faces) = 10 handshakes

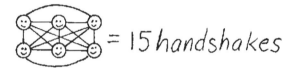 = 15 handshakes

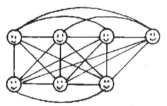 = 21 handshakes

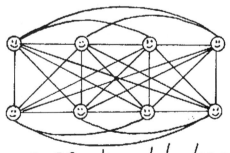 = 28 handshakes

Basic Functions

by James A. Hirleman

Students will organize input and output data in tables to solve a worksheet of function problems. Pairs will also create and submit their own functions for the class to solve.

Materials

One worksheet per student

Introduction

Discussion of Patterns

Write "1, 2, 3, 4" on board. Ask, "What is the next number? How do you know?"
Write "1, 3, 5, 7" on board. Ask "What is the next number?"
"How do you know?"
Continue with these sets of numbers:
"2, 4, 6, 8" (Add 2): "1, 4, 7, 10" (Add 3);
"2, 5, 8, 11" (Add 3); "1, 2, 4, 8, 16" (Multiply by 2);
"1, 3, 9, 27" (Multiply by 3)

Note - It is possible for each pattern to be expressed more than one way; try to find different ways to express them.

Procedure

Introduction to Input and Output

1. "Give me a number and I'll add 1 to it." Students give input, teacher gives output, students record both. Input, Output = (1,2), (2,3) (3,4).
2. "Give me a number, I'll do something to it. You figure out what I did." (Add 4 to each input) Students give input, teacher gives output, students record both. Note - It must be stressed that students should not try to figure out the function until 2 or 3 sets of inputs and outputs have been given. Example: (1,2) could be "adding 1" or "multiplying by 2."

Patterns and Functions

3. Introduce functions with multiplication by doing above with a function of (X 2). Students should be encouraged to make tables. Making tables will help students see patterns in input and output.

Students should leave blank spaces in their tables when an input is missing. They should also try to solve problems by filling in missing input and output and also extend the patterns in the tables.

IN	OUT
1	5
2	6
3	7
5	9

Student Computers

1. Bring one student to the front of the class, give student a card with a function on it. (Multiply by 3) Other students give input, the "computer" gives output, students record data in tables. When students know what the "computer's" function is, they can become the "computer".

2. Repeat 1 above until understood by all.

3. Line up the "computers" side by side. Give input to 1st "computer", 1st "computer" gives output to 2nd "computer", 2nd "computer" gives output to class, eg., 2 —> (+1) —> 3 —> (x5) —> 15 —> (-5) —> 10

Students record data in tables.

(It may be helpful to label each computer initially so that students will be able to see what is happening to their input.)

Worksheet (Pairs optional)

1. Pass out worksheets.

2. Do first problem with students. Give input and output. (1,3), (3,5), (2,4)(, (4,6), (6,8) Students arrange data in tables. (Leave spaces for missing input and output)

3. Students finish worksheet.

   (Inform students that the final problem is a two-step function. Solution — First multiply by 2, then add 3.)

Extension

Each student pair submits 2 functions, a one-step and a two-step function. Each submission must have 5 inputs, 5 outputs, and the solution. Students should make functions as challenging as possible. Post inputs and outputs, let other students work on them when time permits.

General note

This lesson works best if presented in two parts. The first lesson should end after the "Student Computers" activity. The second lesson can start with the activity as a review.

Assessment

Journal entry: When you see a set of inputs and outputs, how do you go about trying to figure out the function (rule)? Give an example.

Patterns and Functions

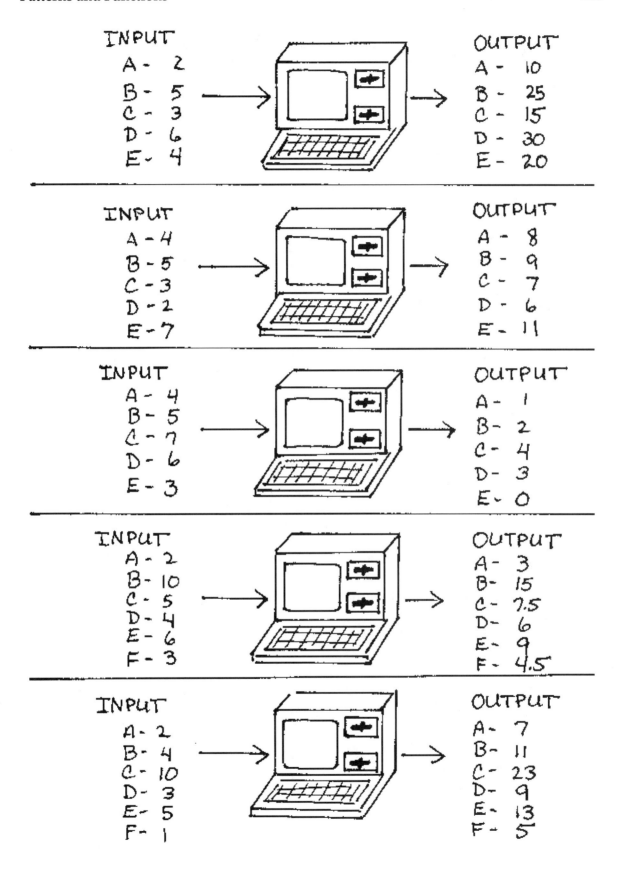

Function Machines

by Penny Hall

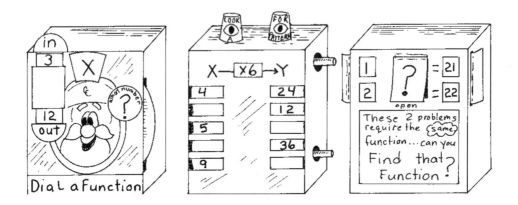

Build your own function machine for use in practicing math facts and recognizing patterns.

Materials

A few days before actually teaching this lesson, announce to our class that they need to bring empty cereal boxes, milk cartons, and toilet paper or paper towel tubes to class. (Have extras saved yourself.) Note: Sample function machines need to be prepared before teaching the lesson. Collect the following materials:

>Empty cereal boxes or 1/2 gallon milk cartons
>paper towel tubes and/or toilet paper tubes
>3" strips of paper taped together to make a long strip
>marking pens
>masking tape, scotch tape, glue, glue sticks
>scissors
>Exacto knife
>tag board or file folders
>crayons
>colorful paper, foil
>drawing compass
>construction paper
>paper fasteners (brads)

Patterns and Functions

Anticipatory set

Show your function machines to the class and watch their reaction! They will hopefully become curious and want to know all about the funny boxes. Once they are curious they will be ready to get into this project with enthusiasm!

INPUT	MODEL	CHECK FOR UNDERSTANDING
Discuss concepts that are needed. (Math facts, addition, subtraction, multiplication, and division.)	Show math problems on the board, working problems backwards and forwards, filling in blanks, such as 6x _ = 18, _ - 4=6, and 5+ _ = 12.	
Explain that this way of presenting and working problems can be called "Function Machine" problems. One number goes into the machine, something happens to it and out comes another number.		Ask if anyone can explain what a function machine does.
Would making your own function machine help you to see how they work?		Students respond.
Let me show you how my machines work. Can anyone guess how the last machine works?	Demonstrate how two of the machines work.	Let a student demonstrate the last machine.
Instruct the students to examine the machines and see how they work. Ask for questions. Tell the students that they can work in pairs, groups, or alone. 1. Share ideas and material. 2. Work quietly and cooperatively. 3. Be creative!		

Exploring

1. Observe the interaction. Notice work procedures. Who is working alone, how are the jobs being divided, who is planning, and who is diving in without a plan?

2. Offer assistance when needed. Some students will request help with the cutting, some may ask for assembly information, and some may get so involved that they forget the math problems for the machine.

3. Provide an extension. If some students have finished early, they may help others, make another different machine, or figure out someone else's function.

Summary

Have students share their machines and tell how they work. Ask about how they made their machine and if it was a good way to do it? What would they have done differently? Have the class present solutions to the different machines. Save the machines in an area of the room for students to examine and use during free time. Make new number rolls for the machines during another math period if the interest is high.

This activity will probably take two or three math periods of 50 minutes each. You can shorten it by eliminating the making of the machines and have the class take turns with the three teacher made machines. Making the machines themselves is highly recommended.

Assessment

Use the machines to asses students' understanding of functions by looking at the problems they make in their machines.

CHAPTER 6
Logic

Guess the Secret Rule A sorting activity K-2
Change One Using attribute blocks 3-5
Guess Two Process of elimination 1-3
Dead Rat A simple strategy game 3-5
Guess My Number Using clues to guess a number 2-6
Color In A strategy game 2-6
Four in a Row A game like tic-tac-toe 4-8
Logical Thinking Maximizing a sum 4-8
Nuts Table logic 5-8
Poison A strategy game 4-8
Matrix Logic Using a matrix to reason 5-8
Logical Algebra A cooperative logic activity 5-8
King Arthur's Problem Using patterns in logic 6-8
The Problem of the 21 Water Casks Hypothetical thinking 5-8

Guess the Secret Rule

by Carolyn E. DeVere

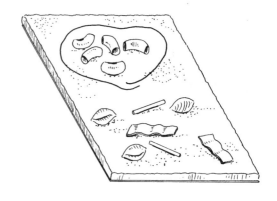

"Guess The Secret Rule" is a concept-level sorting game to be used with small groups of kindergarten students. Invite them to "play a game" at small group tables.

Develop understanding of a set as a collection of items with a common property.
Identify a criterion for membership in an existing set.

Materials

- A piece of carpet
- An 18" hoop or loop of yarn
- A bin of items to be sorted, including: an assortment of small toys (cars, trucks, whistles, plastic animals, etc.); buttons; Unifix cubes; Peabody Kit markers; wooden blocks and beads; colored macaroni, etc.

Introducing

"We are going to play a game called 'Guess The Secret Rule'. I have a secret in mind. It has to do with how I could sort these materials. I will put things in this circle. (Indicate hoop). How do I decide what goes in the circle and what doesn't? That's the secret, for you to guess."

Explain the three rules: 1) Person with secret rule calls on people to guess; 2) There must be five things in the hoop before guessing begins; 3) All the objects in the hoop must "fit" the secret rule.

Model how the game works: Hold up each item as you examine it and decide whether it goes in the hoop or back in the bin. When five items are in the hoop, ask: "What's the secret rule?" Allow each volunteer an opportunity to guess. If no one gives a viable answer, add three more items to the set before inviting the children to guess again. Discuss what an appropriate name for your set might be. Tell the children that now they will take turns showing their secret rule. Give them a moment to think of one.

Logic

Exploring

Have the children take turns thinking of a rule and putting items in the hoop to show the rule. The child who is demonstrating a rule chooses guessers to try to identify his/her rule. The children play the game among themselves, with the teacher prompting only as necessary to keep the game moving and to make sure that each child has a turn.

Summarizing

Review with the children the sets that each child constructed and the name assigned to that set by the group. Tell them they've done a super job.

Assessment

Observe each child as his turn comes to illustrate his rule with five examples in the hoop. Can he show his rule with five examples? Also, consider how many attributes are used in his rule e.g. thick and red. Using more than one attribute in his rule correctly is more advanced.

Change One

by Carolyn Hoffman and Karen Szakacs

Students will identify the four attributes of a block and choose another block that is different in only one way e.g. big, thick, red triangle, and little, thick, red triangle. Students will also demonstrate the ability to explain what attribute was changed e.g. size.

Materials

Attribute blocks

Purpose

The purpose of this lesson is to introduce you to logic so you can figure out what comes next if you follow a certain rule.

Anticipatory set

I am going to play a game with you. I am going to try to trick you! Listen carefully and try to guess my rule.

INPUT	MODEL	CHECK FOR UNDERSTANDING
Introduce the blocks. Talk about the different shapes, colors, sizes, and thicknesses.	Show each attribute as you mention it.	S should be looking at T and sitting quietly.
With a quiet hand, can someone explain to the class how these two blocks are different?	T holds up two slightly different blocks. Wait and allow S to raise hands.	S volunteer with a quiet hand, explaining the differences.
Repeat above exercise at least three times to allow S to gain a better understanding	Same as above.	Same as above.

Logic

INPUT	MODEL	CHECK FOR UNDERSTANDING
of the attributes of the blocks.		
We are going to use these blocks to play the game.	Show or point to the blocks.	S should be listening and eyes should be directed at the blocks on the table.
I am going to put three blocks on this piece of paper. Each block will be different from the one before it in only <u>one</u> way.	Place blocks appropriately.	Same as above.
Hold up a blue block and ask for a quiet hand to tell the T where the blue block should be placed.	Hold up a blue block of your choice.	S should raise hands and give appropriate answer. Have S explain why they chose their answer.
Ask S which block could go next.	Hold up a red block, for example. Start off with blocks that <u>do</u> <u>not</u> go next because they are different in <u>two</u> or <u>three</u> ways.	S should raise hands and wait to be called upon. Have S explain their reasoning.
How many agree with S?	Model thumbs up/down.	S show agree/disagree with thumbs up/down.
Continue until everyone has a turn.		

Assessment

Observe student for attentiveness and perseverance. Can the student complete her turn correctly?

Guess Two

by Marsha Heckert

Process of elimination

Students will use logical thinking processes to problem solve using four different color manipulatives, e.g., playing cards, unifix cubes.

Materials

Enough multi-colored unifix cubes so that each student can have the same 4 colors e.g., red, blue, yellow, green. (At higher grade levels, the number of cubes per student can be increased to 5 or 6.)

One book, or other type of divider, for each pair of students.

Introducing

Show students the four different colored unifix cubes. Cover them over with a piece of paper and hide one in your hand. Ask students to try to guess which color cube is hidden; let students guess until they have the correct color. Tell the students that they are going to be playing a game that is similar to this.

Explain that each of the students will have 4 unifix cubes; your partner will have the same color cubes. You will be trying to figure out which two cubes your partner is hiding.

Students are to place a book (or other type of divider) between them and hide their cubes behind this divider. One student, the "stacker," is to stack two cubes; the other student, the "guesser," is to try to figure out which two are stacked. Stress to the students that it does not matter in which order the cubes are stacked; what matters is that the right two colors are guessed. (** A more difficult version of this game can be play with 3rd or 4th graders where they have to figure out both the correct color and order.) The "stacker" cannot say which cube is right (i.e.., "The red one is right.") The only correct responses to a "guess" are: (1) one color is right, (2) neither color is right or (3) both colors are right, (in which case, the game is over).

Logic

After explaining the game to the students, have a volunteer come up to the front of the room and play the game several times with the teacher. Review the responses with the students, having them repeat them several times.

(This activity is most manageable when taught and played in small groups of 6-8 students arranged in pairs. If one group is composed of an odd number of students, the teacher can also play, ensuring that each student gets to play the game.)

Exploring

Observe student's responses to the "guesser's" attempts. Make sure that the "stacker" is telling the "guesser" if they go one color right. Encourage students to use logical thinking processes to determine the right colors. (Example: "Now if the stacker told you that neither color is right, what does that make you think?" This is not always obvious to younger children.)

Summarizing

Ask students if they noticed any clues that helped them figure out the right colors. Explain what the word "strategy" means and how these types of clues can help you win at games.

Assessment

Even though this game involves using a relatively simple process of elimination to win, many first-graders experience it as a challenging guessing game and enjoy the challenge it provides. By third or fourth-grade everyone in the class will have developed this logical ability. It is interesting to observe the children as they develop their logical thinking while playing the game. They have got it when they can consistently guess the two hidden colors.

Dead Rat

by Pat Marshall

Dan	Natasha
	w
w	
	w

This is a strategy game using cooperative pairs.

Materials

1-inch grid paper for each pair
A pencil for each pair

Anticipatory set

Introducing

Arrange the children in pairs beforehand.
Do you like to win games? Here is a game I learned when I was about your age. I'll teach you to play it. Then I'll let you play it with your partner to see if together you can figure out how a person wins this game.

Describe the game Dead Rat: "I saw a dead rat on the street." "I 1d it."

Next person says, " I 2d it."
First person says, "I 3d it."

And so on until someone says, "I 8 it!" This person loses.

Give jobs

"Figure out who is the older in your pair. That person will start the practice game. The younger one will keep a record of who won." (Model this on the board.)

Guided practice

"We'll do one practice game first to make sure you know how to play. When you are done and the winner is recorded, look at me. (Children play one game.)
"For the next ten minutes play this game taking turns starting the game. If you both think you know how to win, the recorder hands the pencil to the older person and he writes how many games he won with a line under it. Hold up your paper and I'll come over to play a game with you.

Logic

Exploring

Children play games. Teacher monitors the recordings. Pairs that understand the strategy will want to start the game, e.g., "I 1d it."

Extension

For those who have the strategy, I'll change the game by having them choose 1 or 2 numbers each time.

Summarizing

Look at your papers. Do you see a pattern? (Winners should alternate.) How do you win this game? (Start it off.) If you have any younger bother or sisters or friends now you can trick them. Paper monitors collect the papers and return to me.

Assessment

Challenge the student to a game by saying, "I 1d it." If the student protests that she wants to go first, then ask if she can explain why she wants to go first. If she can explain that whoever has the even numbers will lose, then she understands the strategy of the game.

Guess My Number

by Kim Leclaire

Students will use logical thinking to play the game "Guess My Number" correctly three times in each role. The concepts of "more than" and "less than" are used to play the game.

Materials

- A set of ten magnets
- Magnetic board
- Pencil and paper for teacher
- A set of ten unifix cubes for each child
- Pencil and piece of paper for each pair of children

Anticipatory set

Ask if anyone has played the game "Guess My Number." Explain that we are going to learn a new way to play the game.

Introducing

Review " more than" and" less than". Explain that you will pick a number between one and ten and that they will try to guess your number in as few guesses as possible. Write a number on your piece of paper and ask for a guess. (Choose someone with a quiet hand.) Write the guess on the board (e.g. 5) and tell them:

5 is more than my number.

Ask them what they know about your number now. Talk about bigger and smaller numbers.

Get a student to play the game with you at the board. Write a number on your piece of paper and have them guess using the board magnets and saying their number. Use the sentence above to tell them if their number is more or less than your number. They keep guessing until they get it. Play with several students at the board.

Logic

Check for understanding by asking other students why they think the player picked the number they did. Now play the game with a student at their desk, but you take the role of guesser. Make an illogical guess one time and ask the class if anyone can tell you what you did wrong. All hands should go up.

Explain that the person who picks the number is the leader and the other person is the guesser. Tell the students that the leader will hide some cubes and the guesser will guess how many are hiding, writing his guess on a piece of paper. The leader will tell him whether the guess is more or less than his cubes. The guesser will write this down. They will play three times in those roles and then switch for three games. Assign the first leaders, ask for questions and let them begin.

Exploring

Check to make sure all are playing the game fairly and correctly. You may need to remind some of the correct way to say the sentence.

Summarizing

Talk about how many guesses it took to get to the number. Ask if anyone knows ways to get to the number quicker. Review terms.

Assessment

Close-ended question

Give this hypothetical example. Write it on the board as you tell it.

Guess #1: 5   Person says, "Your guess is less than my number"
Guess #2: 9   Person says, "Your guess is greater than my number."
Guess #3: 4

Ask students if "4" is a good guess or a bad one and to explain why they think so.

Open ended question

Ask: Is there a strategy to finding the number in less guesses? If so, how would you cut down on the number of guesses you need to find the number?

Color In

by Nancy J. Woods

The children will use strategy in trying to win a game.

Materials

- Crayons, one per child
- One copy, per pair, of small (4 square), medium (8 square), and large (18 square) grid sheets. (Use 1" square grid)
- Diagram of grid on board

Anticipatory set

What does "strategy" mean?

Do you like games? Do you like to win? Let's play a game and see if we can find out how to win. If we can find a way, then we have found a strategy.

Introduction

Demonstrate on board while explaining rules:

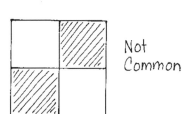

1) Play in pairs with a small (4) grid
2) Each player uses a crayon
3) Take turns coloring one or two squares that share a common side*
4) Last player to color in a square—wins!

Form pairs, write both names on the "4 square" grid sheet and play! Put your initials by the game you win.

Exploring

They play the game in pairs.

Signal

Stop when I say "Strategy."

Logic

Extension

Move on to larger grids so they can practice their strategies. The children can also make their own grids to play at home.

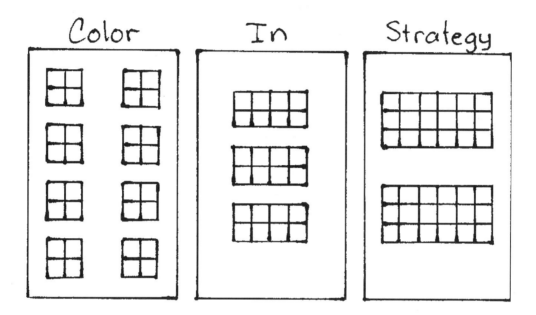

Summarizing

Hopefully the children will see, after practice, the winning strategy. Discuss the various strategies that the children found and then. . .

Assessment

Have children explain what happened when they played the game (who won and how they won). Then ask them, "If you could come up with a strategy for winning every time, what would it be?"

Four in a Row

by Deborah A. Allen

Given graph paper with 10x10 squares and colored markers, students find a strategy for getting four in a row.

By looking for a winning strategy students discover what they can do: they can think critically.

Materials

- 10x10 inch grid for each pair of children
- Inch-square markers (about 50) for each pair
- Overhead transparency of 10x10 grid and markers for the teacher

Anticipatory set

Relate Four In A Row to the game Tic Tac Toe.
Ask:

- How many of you have ever played Tic Tac Toe?
- When you played, did you ever find a way to win or did you just put your marker anywhere?
- Did you ever come up with a "plan" or "method" to win?
- What was your plan?
- Can you think of another word that might mean "plan?" Another word is strategy. (Write strategy on board.)

Instruction

INPUT	MODEL	CHECK FOR UNDERSTANDING
Tell the Rules of Game 1. To win, you need four in a row. 2. A row can be horizontal, vertical, or diagonal.	On transparency demonstrate four squares vertically, horizontally, and diagonally in a row.	Have volunteers explain the game again.

Logic

INPUT	MODEL	CHECK FOR UNDERSTANDING
3. Each player takes turns placing one marker next to any other marker on the board. Play three or four games of Four In A Row with a student using the overhead projector. Tell rules of game.	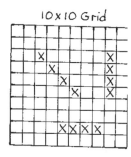	

Guided practice

Play the game with one or two students until everyone understands how to play.

Independent practice

Let students play the game. Ask students independently if they have found a winning strategy. Teacher mingles among groups. Gives help only when necessary.

Sometime during independent practice you may want to let some of the children share their thinking with the whole class.

Students play game again trying to use new strategy.

Summary

- Teacher asks students how many found a winning strategy?
- What was the strategy?
- Use overhead projector and let students show their strategies.
- Why are finding strategies important?
- What are some times you may want to find a strategy?

Assessment

After playing the game over several days, arrange the children in groups of four (moderator, recorder, questioner, praiser) and have the group write down what they learned about how to play a game with a partner (cooperation) and what they learned about winning this particular game (strategy). Then, to assess each individual, interview each one and ask them the same two questions.

Logical Thinking

by Felice Dinsfriend

Decision Making

Given a place value sheet, students will place any given number in the right slot to maximize the sum.

Students will work together as a group to find a successful strategy for maximizing the sum that the entire group will be able to explain.

Materials for each student

- Paper
- Pencil
- A number cube (I used a 5-10 cube from *Real Math*)

Anticipatory set

Today we are going to play a new game using our 5-10 cubes.

Introduction

1. For this game we will first play as a whole class and then break into groups. You will need to have a piece of paper, a pencil, and a 5-10 cube. Set up the following on your piece of paper. (Put the blank addition problem on the board or, if possible, have a sheet with several blank addition problems already made up).

2. Explain that the cube will be rolled five times. If a ten comes up, the cube will be rolled again. Each time pick one of the blank slots on your paper and put the number in. Once you have put the number down it cannot be moved.

3. Explain that the object is to place the numbers so that they will get the largest sum possible when they add.

4. As a class, play the game through once or twice and then break them into groups and give them jobs.

5. Tell them that in their groups they need to play the game through a couple of

Logic

times and then they need to find a strategy that the entire group can follow in playing as a team.

6. When they have found a strategy they need to write it down. Explain that the strategy should be clear to all group members and take in all possibilities.

7. Tell them when the lights are flicked on and off that is their signal to stop playing.

Exploring

Observe the interaction. Circulate and clarify, if needed, to groups where they all have the same confusion. Usually, there are no major problems. Have groups who have found strategies play as a team against each other.

Extension

Add to find the smallest sum.

Subtract to find the largest or smallest difference.

Multiply to find the largest or smallest product.

Summarizing

Ask groups to share their strategies. Put the different strategies on the board and discuss them. (Are they the same or different, etc.). Ask if one of the strategies worked better than the others. If extensions were done, ask if the strategy changes or stays the same.

Assessment

Have students show the addition problems that they generated using the number cube and explain when and why they placed each digit where they did. Then have them compare the sums they obtained this way with the sums they would have obtained by generating <u>all</u> of the numbers first and <u>then</u> placing the digits.

Nuts

by Pat Marshall

Table Logic

Students will use a table to reason out a hypothetical situation.

Materials

Three paper cups, two different kinds of nuts (or beans), paper and pen for each group.

Anticipatory set

Show them three mislabeled cans of nuts and pose this problem to the class:

The nut cannery goofed last Friday. The night crew put the wrong nuts into every can! There are three labels for the cans. The person who filled the nut bins that pour into the cans used the wrong nut bin for each label but he doesn't remember which nuts went where. All he knows is that the cans labeled "peanuts" definitely do not have peanuts in them, the cans labeled "almonds" do not have almonds in them and the cans labeled "peanuts and almonds" definitely do not have a mix of nuts in them. If you could open only one can to see which nuts are where, which one would you open?

Introducing

Model making their own cans and labeling them with a pen.

Make a 3 by 4 chart with **Peanuts, Almonds,** and **Peanuts and Almonds** labels on the three columns. Underneath each label, list what could be inside the can.

Model what you would know if you opened the can labeled "peanuts" by x-ing out what you know is not in each can. If the students are not used to working in groups, model the four jobs that need to be assigned in each group e.g. the chart maker, the can opener, the reader, the questioner. Set a signal for attention and practice it e.g. a bell dings and the students stop talking in groups and face the teacher. Then let students work in groups of four to solve the problem. Give them about 15 minutes.

Exploring

Keep an eye on the interactions as they are setting up the jobs and intervene in any group where there is arguing that does resolve itself within a couple of minutes or that is getting too loud. Once they are settled into problem solving and actually have begun to set up the "can" and to draw the table that they will use, scan the groups for participation. If there is a child who seems not to be participating, make sure he has a job and remind the group of the importance of the job.

To encourage active involvement of all members of the class, it is usually a good idea to let students who speak the same language to work together in a group of four. When you visit the limited English or non-English speaking groups of four, make sure they know the English for "peanuts" and "almonds" and let them practice their explanation with you so that a representative from their group can present their results during the summarizing part of the lesson.

Summarizing

Ask for representatives from the groups to explain which can their group decided on and to demonstrate, using their cans or charts, how they decided on it. When all have had a chance to voice their choice, the teacher can take a can opener and "open" the can that seems to be the most popular choice. Have the whole class say what is in the can as it is poured out. Ask what is in the other cans. Then open and empty the other two cans.

Have the students discuss what went well in their groups, what they liked about group work, and what did not go well in the groups. Did everyone participate? How can each group increase the participation within their group? How did other people's ideas help their thinking? Was anyone afraid to speak up in the group? How can groups be conducted so that no one feels afraid to speak?

Assessment

Metacognition: Students write about their experience in solving this problem. Prompt: What was your first response to this problem and what happened later in your group as you worked on it? What do you feel you got from working on this problem with a group?

Poison

by Ginni Loscuito

Students will play the game Poison with 13 or fewer objects and record results of their winning strategy.

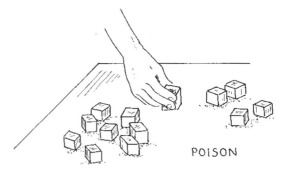

Materials

Thirteen objects (e.g. cubes, lima beans, smooth stones, paper clips)

Anticipatory set

Has anyone ever heard of the mathematical game of Nim? We're going to play a game similar to the game of Nim called Poison.

Introduction

Poison is a version of the mathematical game of Nim. It can be applied to all grade levels and can be of benefit to all ages in teaching logical thinking.

1. This is a game to be played with two people. Thirteen objects are in the center of two people. Each person takes a turn removing 1 or 2 objects from the center until all objects have been removed. The person left with the last object is the person poisoned. The object of the game is not to be poisoned.

2. Play the game at least five times each with the others in your group. Each time you play think of the possible ways you can win. Think of your own strategy for winning and share your ideas with your group. How can you avoid being poisoned? What is a good strategy for winning? Each group will consist of four people; #1 will be the reporter, #2 will be the recorder, #3 will be the materials manager, and #4 will be the rules monitor. The recorder writes down the ideas of the group and will write down the agreed strategy of the group. The reporter will share the group's ideas with the class. The materials manager will get the materials for their group. The rules monitor makes sure everyone is participating and monitors 12" voices.

3. After each group has had a chance to play the game five times with each other, discuss their partial strategies thus far. Have the groups play longer if necessary to try to agree upon a winning strategy. Present each groups strategy on the overhead or front board. Ask for questions.

Exploring

Circulate and observe the students. Encourage groups to stop and examine their strategies if they find themselves stumped. If a group wants to poison the teacher make sure everyone in the group is involved by playing poison with each member of the group.

If a group has found the winning strategy, try an extension of the game by adding to the number of objects and playing the game. For an extension, smaller games (4 or 7) can by played.

Summarizing

Explain that the game of poison has a winning strategy, meaning there will always be a winner and a loser. Winning the game of poison involves logical thinking. Encourage each student to state his own winning strategy.

Assessment

If you know the winning strategy:
To assess if children have the winning strategy, have them challenge you to a game. Put the 13 cubes before them and ask them if they want to go first or second. If they say "second," still try to win even though you are at a disadvantage. They know the strategy if they can beat you.

If you are still somewhere along the learning curve:
To assess, have the students write a three-part journal entry about the game of Poison in which they state how many times they have played the game, how many times they have won and the last part which starts out either "I know I'm going to win if..." or "I know I'm in trouble if..."

Matrix Logic

by Jill Hohenshelt-Veach

Using a matrix to reason

Materials

- Pencil
- Erasers
- Piano Lesson and Family Vacation Problems
- Large Chart of each Problem
- Graph of vacation activities

Anticipatory Set

Does anyone here play an instrument? What kind? Does anyone play the piano? We are going to look at a problem about a family who has three people that play the piano.

Introduction

Problem 1 - The Piano Lesson
Talk about the problem solving strategies that have been covered previously in class. Distribute the Piano lesson problem and put chart on front board (identical to problem).

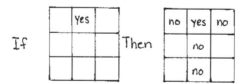

Work problem on the chart explaining how a matrix can be used with clues to solve a problem logically. As students follow along be sure to check for understanding. Model roles students will have in groups. (read, record, check, report).

Stress the importance of each column and row having only one "yes" answer, but many "no" answers possible, ie. one practices at each time, never two at a time.

Model showing that if you have a "yes" then all other squares in that row and column are "no".

Logic

Finishes problem and asks for questions.

Leave the finished chart on the board for the students to refer to when they do the Family Vacation problem.

Problem 2 - The Family Vacation Problem
Ask the children to vote for the activity they like the best, swimming, hiking, camping or fishing. Each student gets one vote. Record the results on a bar graph. Which activity won in the class? This is the activity that we like the best. We are going to find out what activity the Grand family likes to do the best.

Call on student to read the paragraph. Form the children into groups of four. Have the students decide on the roles of their group members - recorder, reader, checker, reporter. These roles have already been modeled by teacher in the piano problem. (If your students have not had much experience in working in cooperative groups, you may need to help some groups to decide.)

Guide student through first clue. Student assists. Ask for question, stressing one" yes" answer for each row and column. Mention that clue #3 is strange. (Rephrases the clue, if necessary.) What activities can we not do if a lake has no water?

Explore

Students work on problem in their group, each student has a job but the group discusses the problem together before coming to any conclusion.

The teacher's job now is to observe the groups. Do not help unless the students are going way off track or if they all have the same question. Gives only enough assistance to get the group thinking again.

Use a prearranged signal to get groups back together as a whole to discuss their thinking.

Summarization

Ask, "What did you find out?"

Ask them to read each clue and tell the class what the clue told them. If only a few students raise their hands, additional discussion may be necessary. A review of the piano problem is useful.

Continue the discussion of the problem until the chart on the board is complete, Allow the students to fill this in when they explain their clue. When a clue has been read and information has been discovered, ask the students if additional information is known without reading another clue, ie., if a "yes" has been discovered then all spaces in that row and column are "no."

When the matrix is complete be sure to finish the problem. There are two questions at the end of the paragraph:

Which activity do they enjoy the most?

Which activities do the other lakes offer?

P.S. Teacher may want to put category names on the students handouts. The children seemed confused with this on my first run through.

Answers

PIANO LESSON
  Betty—9:00
  Bob—3:00
  Brenda—5:00

FAMILY VACATION
  Lake Z—swimming
  Lake F—hiking
  Lake S—fishing
  Lake R—camping

Assessment

Students write a journal entry based on this prompt.

"Sometimes, after gathering pieces of information, we can use logic to find out more information. What did you like or not like about using the matrix? What information did the matrix help you to find that you didn't know at first? How did the matrix help you?"

Source: *Logic, Anyone?* by Beverly Post and Sandra Eads

Logic

PIANO LESSONS

Three children in one family are taking piano lessons. The family has a schedule that gives each child an hour to practice. The practice hours are 3:00, 5:00, and 9:00. Find out when each child practices.

1. Betty practices at either 5:00 or at 9:00.
2. Bob doesn't practice at 5:00.
3. Brenda doesn't practice at 9:00.
4. Betty doesn't practice at 3:00.
5. Bob practices two hours before Brenda.

FAMILY VACATION

The Grand family wants to choose the perfect place for a vacation, but they have some difficulty making the final decision. Finally, they decide on Lake Z because it is the best place to do their favorite activity. Which activity do they enjoy most: fishing, hiking, camping, or swimming? Which activities do the other lakes offer?

1. Lake Z and Lake F have no camping facilities.
2. Lake S does not allow swimming.
3. Lake R has dried up.
4. There are no fish in Lake F or Lake Z.
5. Lake F is known for excellent hiking.

Logical Algebra

by Pat Marshall

A cooperative logic activity

Given a set of six clues each group will use logical language (if, then; and, or; all, some, none) to find the first six numbers in a sequence.

Materials

Poem: **Jabberwocky**, by Lewis Carroll.
An envelope containing six clues and one blank sheet of paper for each of the groups.

Anticipatory set

Recite Jabberwocky by Lewis Carroll, a mathematician famous for writing Alice in Wonderland and through the Looking Glass.

Introducing

Ask the children what the poem is about. How do they know? Most of the words in the poem are nonsense yet it is possible to make some sense of the whole thing. Sometimes it is possible to make sense out of what seems be nonsense. Here is another one. See if you can figure out and an answer to this: I'm thinking of two numbers. The first one is half of the second. The second number is one less than five. What are the two numbers? CFU: Show me each number using your fingers. How did you figure it out? Which one did you figure out first? Then what did you do? (Record on the black board what they say using equations: S=5-1; F=1/2 x 4. In groups of six you will be figuring out six numbers using six clues. The numbers form a special pattern which you will try to guess. Each person in the group will receive one clue which will be read to the whole group but not shown to anyone in the group. Your group's job is to 1. find out the six numbers and 2. figure out a pattern in the numbers and write the seventh. Two special jobs in the groups: Clue Person gathers materials and distributes clues; Recorder write _____ on paper and records guesses and clues from the rest of the group. (Model the jobs with two students.)

Exploring

Children will gather and distribute materials. Then after reading the clues as a group and using logical thinking and a guess and check strategy will attempt to determine the six numbers and a seventh in the series.

Extension

Any groups that finish early will be asked to figure out the next ten numbers in this series. This series is famous!

Summarizing

What are the six numbers? Which number did they find first? How did they get it? Then what did they do to get the second and so forth. Record on the board. What did you guess for the seventh? What was the pattern? (These are Fibonacci numbers: 1,1,2,3,5,8, 13... Each number is the sum of the two numbers preceding it.)

Further activity

Look up Fibonacci in the encyclopedia. Look for Fibonacci numbers in fruit e.g. 3 in bananas, 5 in apples.

Assessment

Metacognition: Students write based on this prompt, "How much did you participate with your group? How did your group do in encouraging everyone to participate?"

Source: This lesson idea comes from the Bay Area Math Project.

LOGICAL ALGEBRA
A COOPERATIVE LOGIC ACTIVITY!
bay area math project

WHAT'S MY PATTERN?

The sixth number is the third number times four and it is the first number times eight

What is the 7th Number in my pattern?

WHAT'S MY PATTERN?

The third number is the second number plus one and it is the fourth number minus one

What is the 7th Number in my pattern?

WHAT'S MY PATTERN?

The fifth number is the third number plus the fourth number

What is the 7th Number in my pattern?

WHAT'S MY PATTERN?

When you add the first six numbers together, the sum is twenty

What is the 7th Number in my pattern?

WHAT'S MY PATTERN?

The third number raised to the third power equals the sixth number

What is the 7th Number in my pattern?

WHAT'S MY PATTERN?

The first and second numbers are the same

What is the 7th Number in my pattern?

King Arthur's Problem

by Deborah M. Engelhart

Making it simple
Looking for patterns

Materials

- 32 objects
- A piece of paper and a pencil for each group of 4 students

Anticipatory set

Today we are going to play a game using numbers. Review: What's My Number Game from the previous day.

Introduction - King Arthur's problem

This is a fictionalized historical problem. King Arthur wanted to decide who was the fittest to marry his daughter, and chose this method. When all his knights were seated at the round table, he entered the room, pointed to one knight, and said: "You live." The knight seated next wasn't so fortunate. "You die." said King Arthur, chopping off his head. To the third knight he said: "You live," and to the fourth, he said: "You die," chopping off his head. He continued doing this around and around the circle, chopping off the head of every other living knight, until just one was left. This remaining knight got to marry the daughter, but, as the legend goes, he was never quite the same again.

1. Read the problem to the class and show it on the overhead.

You will play the game in groups of four. Each group will have a pencil, a piece of paper and 32 cubes. Place the objects on the paper in a circle and number them 1 to 32. One person is to mark off the objects while the other 3 watch and try to figure out which will win. Everyone will have a turn. What you want to do is eliminate all but one cube in the manner discussed.

2. **Show a sample problem using ten objects and do a complete game.**

3. **Play four games so that each person gets a chance to eliminate cubes.**

I want you to find out which cube is always going to win. For the first game use four cubes, use eight cubes for the second game, for the third game use 16 cubes and use 32 cubes for the fourth game. Then try some other numbers. Look for a pattern. Are there any questions?

4. **Assign group membership and jobs.**

Exploring

Pass out materials, then circulate and observe. See if they can predict which cube will always win no matter how many there are. Watch for groups that are spinning their wheels, verbally discussing the problem but never getting anywhere. Have them use real objects or drawings or a table to make it concrete.

Extension

After they have played using the four different numbers, give them different numbers to try to see if the strategy is the same, for example, for an odd number of knights.

Summarizing

Show the problem on the overhead using four, then eight, then sixteen objects.

Explain that there is a winning strategy and if you know what it is you can always win. Discuss what they have noticed so far. "Do you see any patterns? What numbers seem to win?" List the ideas they suggest as possible winning strategies. Challenge the class to test them out to see if any of them work.

Assessment

Journal entry: What did you notice in looking for a winning strategy? What was the best place to sit?

The Problem of the 21 Water Casks

by Jim Casey

Logical thinking using If...then, and/or

Students will test out several hypotheses as how to divide up the water and the casks by using drawings.

Materials

Transparency of the worksheet
Worksheet for each student/group with grouping of barrels; 7 full, 7 half full, 7 empty. —OR— Manipulatives can be used (i.e., small color-coded paper cups or construction paper picture of casks: blue for full, red for half full, white for empty.)

Anticipatory set

Flash worksheet of the 21 casks on the overhead screen and wait for quiet.

Introducing

1. **Present or review concepts.** The "Problem of the 21 Water Casks" is an activity to solve the puzzle of how to divide up the water supply and the casks amongst the homeward bound travelers. Each traveler must leave with an equal portion of water and equal number of casks.

2. **Pose a part of the problem or a similar but smaller problem.** For instance, if there were two travelers and two full casks, two half full casks, and the two empty casks, how would you divide the casks so that each traveler had an equal number of casks and equal quantity of water?

3. **Present the problem to be solved.** (Give these directions): Each team will have a worksheet or 21 "casks" or 21 appropriately labeled mock "casks." Remember, of the 21 casks, 7 are full, 7 are half full, and 7 are empty and must be divided evenly so that each of the three travelers has an equal number of casks and equal quantity of water. If the team can come up with a formula, equation, or drawing that will explain he solution, write down statements that describe your thinking. If a team thinks they have a solution, raise your hands and I'll review your

Logic

answer. Later each group will choose a person to present to the class the group's solution and strategy for solving "The Problem of the 21 Casks."

4. **Discuss the results.** Ask for questions about the demonstration practice problem. Emphasize group effort and that there is no "one way" to find the solution. Each member of the team should have some input — all ideas must be explored as a team. The group must answer their own questions without teacher assistance unless they want a solution checked or are at a loss on how to start.

Demonstrate a signal that will stop group activity and focus attention on the teacher to clarify, discuss, or summarize.

Exploring

The teacher should circulate and observe. Be aware of any team that may have too dominant an individual or, on the other hand, a student who is passively looking on without contributing. The task is a "group-think" activity. Reinforce that point. If any one group is "stuck," ask leading questions or suggest a beginning thought to stimulate the team in a certain direction. Any team that has determined the solution and checked with the teacher may be offered the first of several extensions of the cask problem. Tteams that finish before others are encouraged not to share their solution/strategy until the teams come together to summarize.

Extensions

1. Rearrange the problem. Divide the water equally and disregard winding up with an equal number of the casks among the travelers.
2. Have the students make up their own problems similar to the "21 casks" (i.e., Changing the number of casks or number of travelers).
3. Change the number of trips it would take to transport the water.
4. Restate the problem as a syllogism: "If the first traveler can carry twice as much as the second traveler but only half of what the third can carry, how much is each able to transport?" The number of casks and quantity of water in each can be adjusted to the problem.

Summarizing

At this point the group representatives will present their solutions and strategies. The students (raising hands) or teacher may ask questions to clarify any part of the group's presentation. The teacher will record on the board statements that summarize each team's strategy. Even if there are duplicate ideas on how to solve the puzzle, each team should have their representative take a turn at describing the team's process of coming to a solution. They may use the blackboard or the overhead projector with an overlay of the casks. Manipulatives (mock casks) may be used on the overhead projector to demonstrate the solution. Emphasize that there may only be one solution (Each traveler carries 7 casks — full, half full, empty — containing a total of 2 1/2 casks of water), but there are many different ways to arrive at this answer. There is no "one right way" to solve the problem.

The teacher may also share any strategy that has come up from past lessons (drawings, formulas, equations, or any applicable strategy). Discuss what made the problem difficult. For instance, "Dividing the water without being able to open the casks must have been tricky." Note the number of "variables" or conditions that had to be met and how addressing them one-at-a-time is a good strategy. "Let's find out how much water we have altogether then how to divide it amongst the travelers." Also useful is paraphrasing the problem and asking, "What are we trying to accomplish?" A good summation would include dealing with the quantity of water and the number of barrels.

Assessment

What did you get for an answer? How sure are you that your answer is right?(Rate yourself somewhere between 0 (not sure) and 10(positive). Why do you feel that way?

THE PROBLEM OF THE 21 WATER CASKS.

As a caravan had to cross the desert to return home, each traveler needed to carry a water supply and return the empty casks already used on the journey. The total number of identical casks was 21.

7 casks full of water
7 casks half full of water
7 casks that were empty

The travelers wish to divide the number of casks and quantity of water evenly. The "cask" solution is obvious; each is to carry 7 casks. The difficulty is dividing the <u>water supply</u> evenly without opening up the casks--leaving them just as they are--full, half full, or empty. How can this be done?

BELOW ARE 7 FULL CASKS, 7 HALF FULL CASKS, 7 EMPTY CASKS

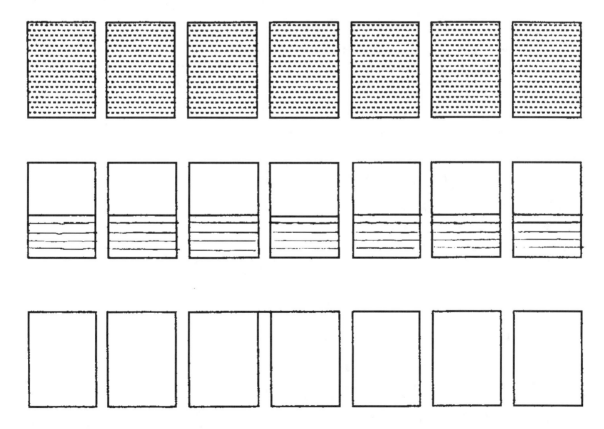

CHAPTER 7
Number

Beginning Number Concepts
Unifix Trains of 1-6   Numbers 1-6 concretely K-2
Place Value
Counting by Tens with Kip   Unifix cubes for ones and tens 1-3
Flats, Longs & Units   Base-10 mat and materials 3-6
Clear The Board   Place value and regrouping 3-6
$1 000 Bill Game   Place value or money denomination 3-6
Basic Facts (add/subt; mult/div)
Sums of Seven   A card game 1-3
Help for Walter   Arranging sums of ten 1-2
Subtracting Numbers Less Than Ten   Acting and writing stories 1-4
"Counting Up" with the Cover-Up Game   Higher decade combinations 3-5
Colliding Cubes   Sums of 11 to 20 2-5
Missing Addends in The Little Red Hen   Matching an equation to a story 1-4
Tally Marks   Multiplication as repeated addition 3-6
Spatial Multiplication with Color Tiles   Making rectangular arrays 3-6
LCM   Solving problems using least common multiple 4-8
Dividing a Number by Itself   Writing story problems with quotients of 1 3-6
Word Problems in Multiplication and Division   Writing about solutions 4-8
Algorithms
Explaining + and - Situations with Base 10 Materials   2-digit problems 3-6
Using the Hundreds Chart to Add & Subtract   2-digit problems 4-8
Showing Partial Products Using Rectangles   Constructing products 5-8
Dividing Money from a Cash Drawer   Short division concretely 4-8
Hundreds of Dollars in the Cash Drawer   Division with 3-digit dividends 4-8
Fractions and Decimals
Fraction Freddy   Naming fractional parts of groups of objects 4-8
The Doorbell Mystery   Dividing a group into fractional parts 4-8
Equivalent Fractions   Using pattern blocks 4-8
Geoboard Fractions   Naming and comparing equivalencies 4-8
Problem Solving with Fractions   A measurement model 5-8
Fraction Strips   Adding and subtracting fractions 4-8
It All Adds Up!   Adding halves, thirds, fourths, sixths and twelfths 5-8
"Tenths" Using Base Ten Blocks   Tenths and hundredths concretely 5-8
Multiplying Decimals   Students present several models 6-8
Percents, Ratios, Proportion
Fractions as Rates   Generating a series based on a rate 5-8
Ratios - Cubes to Tiles   Ratio as a fraction 5-8
Introducing Percents   Several concrete models 5-8
Fun With Percents   Visual concept of percent 5-8

Number

Unifix Trains of 1-6

by Jennifer Moldrem

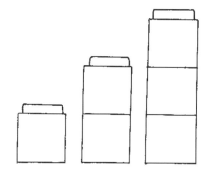

Develop basic concepts of the numbers 1-6 in physical form.

Introduce the number combinations that add up to the numbers 1 - 6 using a hands-on approach.

Materials

Unifix cubes for each student in pre-set "trains": 1 red, 2 blue, and 3 green. Also corresponding ditto to record their findings (see attached).

Transitional activities

The children are being introduced to "adding" with math stories read orally and finger games. They have already been introduced to unifix cubes.

Classroom environment

Groups of six children per table group. Bins of unifix cubes are at each table group.

Noise level will be moderate as they are working with manipulatives and each other. They may talk but must complete their own paper.

Anticipatory set

Put the cubes on your fingers and talk about the differences in the three trains (number and color). Have them make six trains (two of each) and set them in a row on their desks.

Purpose

Tell the children that we will be making the numbers 1-6 using only these six unbreakable trains. Also explain that it will be fun showing the cubes on their fingers and how to make them than just using the numbers (symbols).

INPUT	MODELING	CHECK FOR UNDERSTANDING
We do all the problems concretely. This is one way to show 2. Can you show me a different way?	I hold up one red cube on index finger of each hand.	They hold up blue train.
I ask them to show me with trains how we can make the number 3, etc.		They show all the trains that make 3, etc.

Guided practice

We "play" practice through rest of numbers. Then I show how we can record our answers with crayons on paper. For example, the first one shows how many? (2) Take your red, blue and green crayons and color in two trains that make two. (They color.) Now take turns and "read" your trains to your partner. (One and one is two.) (Faster children who finish early can help slower children. This fosters teamwork. One usually hyper child in my class made a very good peer tutor.)

Independent practice

The students who are capable or ready will work ahead on ditto without needing help. They can then help you help others. Very helpful!

Closure

We restate how we make numbers together with the 3 trains in a game-like fashion once again.

Assessment

Ask children to make a specific number by themselves. As you see which train they reach for first, student thought processes become very visible.

Number

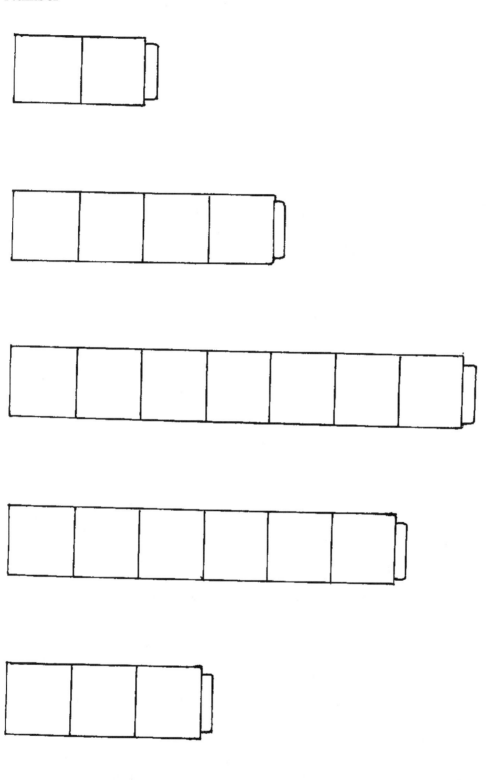

Counting by Tens with Kip

by Jim Cordellos

This is an indirect and informal introduction to the concept of place value as we develop counting by tens. I have no intention of even mentioning the words "place value," but some of the materials I will use could also be used in a place value lesson. This activity will give them a visual of what the numbers represent, thus strengthening their concept of place value, while giving them an introduction to base ten manipulatives.

Materials

First we need Unifix cubes, enough for twenty for each child. One Koala puppet to get the children interested and keep them interested. I will also have a sheet with the numbers 10, 20, 30...100 for them to follow along when we count. Lastly, we need an ample number of Cheerios for Kip, the Counting Koala to share with the children when we are done.

Transitional activities

The students have practiced their counting and have done work in grouping items in groups of tens. But they lack even a crude concept of place value which I think inhibits their ability to judge at a glance which two digit numbers are greater than others. For example, many children cannot say which number is larger, 20 or 30. The long stick will provide the visual for counting by tens that the individual cubes provide when counting by ones.

Classroom environment

This is a small group activity, 4 or 5 students per group. Being that it is one of their first introductions to manipulatives as a learning tool and counting by tens, attention must be paid to each student's understanding of the concept. It will begin as a whole group activity with a lot of modeling on my part and we will move to individual work when I want to check for a specific individual's understanding. There will more than likely be distractions so I will need to maintain their attention with my Koala.

Anticipatory set

I will get them focused by introducing my friend Kip, the Counting Koala. I will explain that it is important for him to be able to count and recognize groups of leaves and twigs that he must provide for this family to eat. He has to make sure he has enough for everyone. I will also add that Kip is shy and scares easily so we must be very quiet around him. I will keep him under my jacket until I am ready to begin.

Purpose

Explain that counting of any kind will be important every day for the rest of their lives. I will ask them where they use counting in their everyday activities. "You count how many minutes are left before your favorite Barney cartoon starts; you count how many points your little league team scores in a game; you count how much money you got for a tooth you lost, etc.

INPUT	MODELING	CHECK FOR UNDERSTANDING
First I will give them Unifix Cubes already connected into groups of ten and tell them to take them apart and count them. "What did you get?"	Kip and I will take apart a Unifix stick of ten and count the cubes.	Have them count each others' while I observe. "Count your neighbor's cubes."
		Chorally: "10!"
Then I will give them the cube sticks with ten cubes and ask each of them to count the number of cubes on the stick as Kip and I count.	Kip and I will count.	I will ask them to follow along and tell how many cubes are on each stick. If they agree that there are ten, show thumbs up.
When we have all established that there are ten beans on each stick I will ask them what the difference is between the two groups. The point is to establish that 10 individual cubes are the same as the long stick.		If we agree that the long sticks have the same number of cubes as the groups of ten that they made, we will give the thumbs up.

INPUT	MODELING	CHECK FOR UNDERSTANDING
I will tell them that we count by ones when we are counting the individual cubes and by tens when we are counting the long cube sticks.	I count some cubes by ones and then count some cube sticks by tens.	I ask them to do it again with me.
I want to show a number with the cubes.	I will start by representing a single digit number and asking them what the number is. If they identify it correctly, I will ask them to show me the number as well.	They show number using cubes.
I want to show a bigger number with the cubes. I will choose a multiple of ten being that I want them to get the concept of counting by tens.	Then I will demonstrate how to represent a two digit number, that is a multiple of ten, using the long cube sticks and ask them what the number is. If they identify it correctly, I will ask them to show me the number as well.	They show number using cube sticks.

Guided practice

I will observe how they do with some other numbers I give them to represent with the cube sticks. Do a couple with them. "Guide" them (in preparation for independent practice if there is one). OR You may only do guided practice.

Closure/Assessment

I will tell them that they did such good work that Kip has some Cheerios that he wants to share with them but we must first put them in groups of ten and count them by tens together. After we count together, I will assess their understanding by asking them to go in order around the table and show me a number on the cube sticks, e.g. 10, 20, 30....

Flats, Longs & Units

by Kari Pfeiffer

Students will use base ten materials to demonstrate an understanding of place value (ones, tens, hundreds).

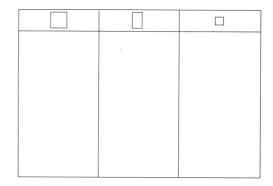

Materials

Base ten materials (flats, longs, units)
Place value mats (appendix)
Spinners (1-9) (appendix)

Transitional activities

This lesson will serve as an introduction to adding two- and three-digit numbers and to regrouping.

Classroom environment

Students will be expected to listen carefully while instructions are being given or while other students are sharing. There will be instances throughout the lesson when students will work in small groups (4). At this time, "twelve inch voices" will be allowed.

Anticipatory set

Ask students to count by tens to one hundred together. Ask them to think of examples of things that come in tens (fingers, toes, players on a basketball court at a time, pennies in a dime, dimes in a dollar, the number of minutes in recess).

Tell them that in our number system, 10 plays a special role and that today we will be playing with the number ten and different combinations of the number ten.

Purpose

Inform students that a good understanding of the number ten will make it possible for them to solve complex (for them anyway) math problems such as 56 + 44 = ___.

It will help them to find out how much money they need to have if they want to buy a 27 cent jaw breaker and a 73 cent chocolate bar.

Procedure (input, modeling, checking for understanding)

1. Pass out some "units" to the class; explain that we will call these "units."
2. Ask students to tell their neighbor how many units they have on their desk.
3. Pass out a few longs to each student. Tell them we will call these "longs."
4. Ask them to count how many units there are in a long and hold up fingers to show it.
5. Pass out a flat to each student. Have them say how many longs there are in a flat. Ask them if there is a quick way to tell how many units there are in a flat (elicit the idea of counting by tens.)
6. Tell students that we can call the pieces of our base-10 set by other names as well. Have them think of names that tell about how many things there are. (Ones, tens and hundreds).
7. Check for understanding by asking students to "Show me 10; show me 100; show me 1, with their flats, longs and units."
8. Now ask students to figure out how they would show me 2, 3, 7. (Have them hold the "numbers" up.)
9. Then, move into double digit numbers and have students show (hold up) other numbers (20, 30, 11, 32, 44...)
10. Finally, move to three digit numbers (100, 110, 120, 121, 122, 200, 210...)
11. Put some "numbers" on the overhead. Ask students, "How many ones...How many tens...How many hundreds?" and "What is the number?" Don't always put them in order.

Guided practice

Introduce *Race to a Flat* game:
Students work in pairs. They alternate spinning a 1 - 9 spinner and placing that number of units on their FLU boards. Explain to students that they should make exchanges for a long whenever possible. The first student to exchange 10 longs for a flat is the winner.

Before beginning, model how the game is played on the overhead, and do a few rounds together. For example, have them discover what to do when we have 8 units on the board and we spin a 4. Ask them, "What are we going to do with all these ones? Isn't it going to be hard to tell if we reach 100? What can we do so that it makes it easier for us to tell how many we have on our board? Make exchanges.

Independent practice

Race to a Flat serves as in-class independent practice.

Closure

Have students explain what numbers help them reach 100 more quickly. Ask them if they were doing math when they played the game. Ask them to explain what they did and <u>what</u> they thought about. e.g. What did you do when...
                      You had 8 units and you spun 3?
                      You had 2 longs 5 units and you spun 5?
Base questions on their particular game and let them draw it on the board or overhead.

Follow-up activities

Allow students to play *"Race to a Flat"* for several days until everyone has "won" at least once.

Assessment

Observe student as he performs the following:
1. Give child 24 counters and ask him to put them in groups of 10.
    Ask: a) How many groups of ten do you have?
                b) How many extras?
                c) How many do you have altogether without counting them one by one?

2. Dictate several numbers (i.e. 107, 12, 230, 325, 17). Have the student show you the numbers with their blocks.

3. Give the child three numerals on small cards. Ask him or her to make the largest number possible from arranging these three numerals. Ask for the smallest. (Symbolic)

Note

I assessed one student after this lesson and found that she could do all three of the activities even when I included a four-digit number. She did have some difficulty, though, in reading the numbers, particularly when a zero was in the tens or hundreds place of the four-digit number.

Clear the Board

by Hallie Atkinson

To develop the concept of place value and re-grouping. This game is harder then the *Race to a Flat* game in **Flats, Longs & Units**. It starts with a flat on the board. The number spun is removed from the board rather than added. Does the student know what to do when she starts the game and spins a 1? When she has some longs and units and spins a 9?

Materials

Flats, longs, units, spinners, place value mats, transparency of place value mats (see appendix).

Transitional activities

During the introduction I will explain to the students that this is an extension of the single-column addition, which they have been doing, to double-column addition.

Classroom environment

Groups of 2, noise level should be moderate.

Anticipatory set

Count by 10s. Identify manipulatives.

Purpose

To give the students the understanding they need to add and subtract "big numbers".

Guided practice

I will model the first activity by having the children give me a two-digit number and I will "show" them the number with the manipulatives. I will do this several times until their faces reflect understanding. I will then have them model several numbers. Once the children are consistently accurate with their manipulatives, I will have two volunteers come to the front of the class where I have a large spinner

Number 291

and a place value mat transparency on the overhead. The volunteers will demonstrate several spins to the class "regrouping" and removing the given amount of units each time, the object of the game being to be first to "clear the board."

Independent practice

The children will pair up and begin play. As I walk around the room, I will ask different teams to explain their strategy to me, and prompt them to talk about the need to re-group. For those teams finishing first, I will suggest they play a rematch.

Assessment

(Same as **Race for a Flat**.)

$1000 Bills game

by Kelli Burns

Place value or money denomination

Develop place value using 1's, 10's, 100's and 1000's.

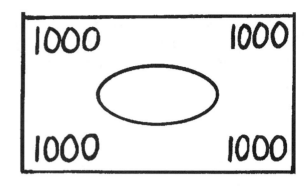

Materials

$1000 Bill Game board, play money (enclosed), place markers, and two dice for each team of two to four players.

Transitional activities

This is similar to the F.L.U. or base-ten materials that they used a few weeks prior to this lesson. Remind them how they added more units to what they had and when they got ten, they traded or regrouped whenever they could.

Classroom environment

Noise level during instructional time should be low. During play time the noise level will be medium to high. Groups will consist of three students.

Anticipatory set

Talk about garage sales. Who has had garage sales? People see prices. Prices are marked. Buyer and seller must be able to count money. Seller collects money.

Purpose

During this game they will learn how to change money to bigger denomination bills. This activity in place value, with an emphasis on money, will help them when they go to the bank or the store.

Number

INPUT	MODELING	CHECK FOR UNDERSTANDING
1. Describe rules of game aloud to students. (See Independent Practice.)	Hold up each item as I describe it. Show movement and actions as I describe them.	Ask each student to identify a few of the different bills. Make sure all students understand rules by asking questions and looking at faces. Follow-up on any questioning looks.
2. Describe exchanging and adding techniques as review, e.g., $210 + $31 Count up from $210 by 10s and then by 1: $210, 220, 230, 240, 241.	Show examples of right and wrong change making. Discuss why it is right or wrong. Everyone should be satisfied that the adding is correct. If not, recount.	As examples are shown, ask children whether it's right or wrong. Have them respond with thumbs up or thumbs down.
3. Ask children how they know when someone is the winner. Tell them to respond outloud without their hands up.		Listen to call outs to make sure all children say, " When someone gets a $1000 bill!"
4. Ask if there are any questions or if every one understands how to play.		Answer all questions
5. Assign groups and instruct how to arrange materials.	Set up a board properly for children to see as they set up their own.	Check each board for proper set up.

Independent practice

Directions for the game: Each player has a place marker. In turn, each rolls the dice and moves the spaces indicated. When landing on a square, the player announces the dollar value which is either the amount shown on the dice or a multiple of it, e.g. "amount x 3" means take three times the amount shown on the dice. Then they collect that amount from the bank in the center of the board. When they collect their money, they should start with the highest denomination possible and work their way down. They are counting aloud to the other players as they do this. They then place the bills in their own pile of money gathered. Whenever possible, the players redeem their money for the next higher denomination. The first player to get a $1000 bill wins.

Respond only to student questions that no one in the group can answer. Provide lots of student praising.

Closure

After putting away materials, ask children about what they learned during the game. What was hard? What was easy? Discuss.

Assessment

Ask a student to show you the way to change up or trade in bills for higher denominations.

Number

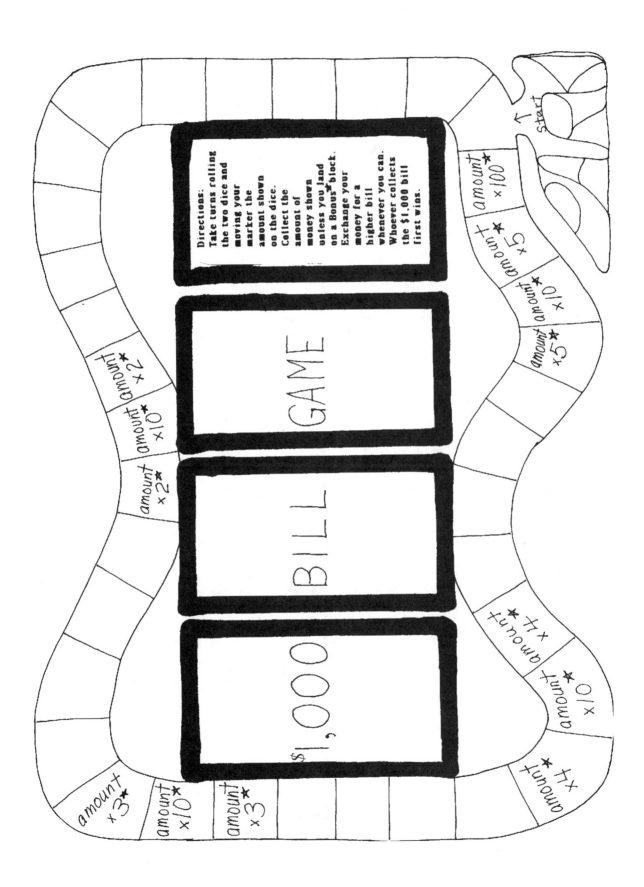

296 Math Plans

Number

10	10	10	10
10	10	10	10
10	10	10	10
10	10	10	10
10	10	10	10
10	10	10	10

Math Plans

Number

Sums of Seven

by Darcy Cooper

Making pairs of playing cards showing a sum of seven pictures (hearts, clubs, diamonds, or spades)

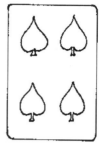 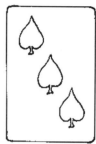

I wanted the children to understand that different number combinations can add up to the same total. The game I played with my class was called "7s", so the children will learn that not only do 3 hearts + 4 hearts = 7 hearts, but also 2 hearts + 5 hearts, 6 + 1, and 7 + 0 hearts equal seven.

Materials

Each pair of children needs two decks of cards put together with all the cards of value greater than 7 taken out. Jacks are worth zero. Other face cards can be taken out.

Transitional activities

The children have been learning to play math games with cards and most of their math lessons for the past two weeks have been centered around these card games.

Classroom environment

Small groups of two played together. There was a certain degree of noise because the children had to talk to each other while they were playing the game but this is a common occurrence in my classroom.

Anticipatory set

"O.K., is everyone ready to learn a new card game? First I would like everyone to take your deck of cards and divide it into two piles: one pile for the cards higher than 7 and one for the cards 7 and lower." (I had them divide the cards themselves because four of the children are English-as-a-Second-Language (ESL) students and I wanted them to have practice with the concept of greater than and less than.

Purpose

To practice combinations of numbers that add up to 7.

Input

We are going to play a card game called "7s." Everyone pick a partner and place your deck of cards face down in front of you. Now, each person lay down 2 cards in front of you. The first person takes the card off the top of his/her pile. If the person's card and one of the cards on the table add up to 7, then the player takes both cards. I want you to count the number of hearts, spades, clubs or diamonds that are on each card. (Few of the children recognize that the number in the corner of the card is the same as the number of objects on the card). Then the second player takes a turn and does the same thing. If the player does not make a match then he discards his card on the bottom of the deck. Whoever has the most cards at the end of the game wins.

Modeling

As I am explaining the game I show them how to do it with the cards. I draw one card off of the top of a pile, count the objects, and then add it to each card on the table until I get 7 - if I can.

Check for understanding

While I am doing the modeling I ask the children to help me add, and ask them step by step, "What am I supposed to do now?" This way I can see if they understand the directions, as well as the idea behind the game.

Guided practice

After my explanation I will watch the children play a couple of times with me talking them through it. I have to remind many of the students to put their fingers on each object on the card when they are counting because many of them have trouble with one-to-one correspondence.

Independent practice/Assessment

I watch the children play without saying anything unless they ask for my help. See if any cannot get more than one correct combination for 7.

Closure

After the game ends I ask the children, "What different combinations of numbers add up to 7?" With their input we figure out all of the combinations. I write them as equations on the board e.g. $1 + 6 = 7$.

Help for Walter

by Michelle Richter and Tami King

Using manipulatives (graham cracker bears) for counting, students will write down different "sum equations" of 10.

Materials

A stuffed bear (Walter)
Ten "bears" per student
Two worksheets per student

Anticipatory set

"I brought a friend with me today who has a problem and needs your help." Go on to tell the class the story about "Walter" (the bear).

Story

Walter's owner's name is 'Katie'. Every night before Katie goes to bed she arranges her ten bears the same way on her two shelves. She always puts six bears on the top shelf and four bears on the bottom shelf. The problem is that the bears are bored! They want to sit by some of their other friends, but Katie arranges them on the shelves the same way every night. Walter needs your help to find out some different ways to arrange the ten bears on the two shelves.

Purpose

This lesson will give the students more opportunity to write equations and reinforce the concept of addition through a 'real life' situation. This lesson can also be an introduction to the concept of combinations.

Instruction

1. **Input:** "So Katie has ten bears and she always puts four on the bottom shelf and six on the top shelf."
Model: Draw shelves on the board or overhead projector and put 4 bears (use crackers on overhead) on bottom and 6 on top.
Check for Understanding (CFU): "Do you see how Katie always puts her bears?"

2. **Input:** "You will each get your own shelves and 10 bears. We'll all see how we can help Walter. Please do not eat your bears. We all will eat them after we help Walter solve his problem."
CFU: Ask some students: "When will we eat the bears?"

3. **Input:** "Instead of always having 4 bears on bottom and 6 bears on top, how else can Katie arrange her bears on the shelves? Remember to use all 10 bears because Katie loves all of them."
CFU: Give students time to manipulate the bears on the shelves before you call for responses.

4. **Input:** "Can someone give me another way Katie can arrange her bears - thumbs up."
CFU: Choose a volunteer from those with thumbs up.

5. **Input:** "Let's write this equation on our 'Help for Walter' paper, because this is another way that Katie could arrange her bears."
Model: Write it on your paper that is taped to the board.
CFU: "Put your hand on your nose if you've written this equation down."
Do steps 4 and 5 several times then, as exploring, let them work in pairs - one arranging bears and the other recording the equation. Take turns. As a class, go over all the possibilities.

Summary

Ask - "How did we help Walter?"
Eat the bears and mention how math is involved in many day-to-day real life problems.

Closing comments

This lesson was given in a first grade classroom and it worked perfectly. The children loved it and they were just at the point where this type of lesson would mean the most to them. The lesson's great appeal is the use of a bear puppet in the telling of the story.

Assessment

Read the equations they wrote. Did they find most of the combinations for 10?

Number

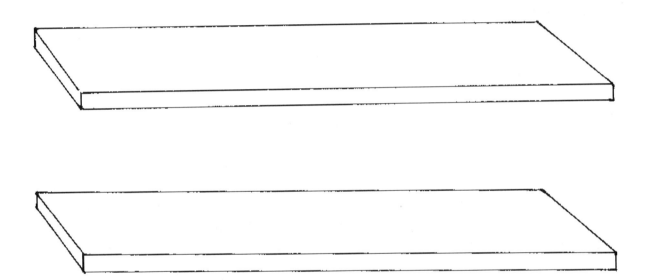

Subtracting Numbers Less Than Ten

by Monika Kuester

To conceptually teach a second grade class how to subtract using numbers 1 - 9.

Materials

Unifix cubes
Paper and pencil

Transitional activities

The second graders have been trying to do subtraction problems symbolically with great difficulty and it is fairly obvious they are not grasping the basic concept. Even when using a number line they seem to become confused as to whether the answer is the number of spaces they have moved or the number they landed on. I feel the basic concept needs to be reinforced before they can go on.

Classroom environment

A low noise level is expected. Although this is a whole group lesson led by the teacher, some discussion among neighbors will be encouraged.

* Before actually beginning this lesson, hand out pre-counted piles of unifix cubes to the children 5 to 10 minutes before recess time to allow them to "explore" with them so when we come in from recess they will be all ready to begin the lesson. Explain that after recess we are going to use them to help us learn.

Anticipatory set

Discuss what subtraction is, to take away, and with the children come up with a reason why understanding the concept of subtraction is important. Brainstorm some ways that they use subtraction in their everyday lives. Give some relevant examples, e.g. 5 people in family and 1 is sick so we subtract one of everything when we set the table: 4 plates, 4 glasses, 4 forks, 4 napkins, etc.

Purpose

To gain a fuller understanding of the basic concept of subtraction. They now know that this concept is beneficial to them because examples of how this concept is used in their every day lives have been discussed.

Input and modeling

Instruct children to take four unifix cubes that are the same color from their pile and snap them together. Next ask them to hold their towers up. Reinforce that they now have a tower of four. Count the number of cubes together. Now explain through demonstration that if we take this tower of four and take away one cube we will have a tower of three. Count this tower with them. Guide the children through this step as they do it along with you. Next ask them what their towers will look like if we take away two more cubes. Continue this procedure by again having the students make a tower this time of nine cubes and take two away. What would we have if we took five away? After several examples and seeing that the children are grasping the concept I would let the children get into partners and ask each other different subtraction problems.

Check for understanding

I would constantly be checking for understanding in this lesson by having the children hold their towers up or also holding up the answer they came up with on their fingers.

Independent practice

Having the children working in pairs will give me a chance to quickly go around the room to look over the children's shoulders as they work with each other and to overhear some of the discussion that is happening. * Activity: Pretend the cubes are something else e.g. little cars, little people, etc. (emphasize little things) and act out a subtracting story for your partner. Then have your partner write your story.

Each partner gets to make up a story so there are two papers for each pair. Model using this story form: "There were 5 puppies. 2 of the puppies left with new owners. Then there were 3 left. By Alan."

Use this story form:

```
There were _____ _____.
_____ of the _____ _____.
Then there were ___ left.
            by _____
```

Closure

In closure to this lesson, ask for volunteers of pairs to come up in front, demonstrate a problem and read the stories they wrote with their partner. **CFU:** Whispering the answer outloud is one method I plan on using along with possibly calling on one student for an answer and having the rest of the students agree or disagree by giving me a thumbs up or down. Again ask students why they think that this concept is so important and if they can relate some situations in their everyday lives where they plan on using this concept. I gave the example of how I use subtraction every day to keep track of my money in my checkbook.

Assessment

Read the stories to see if they make sense.

"Counting Up" with the Cover-Up Game

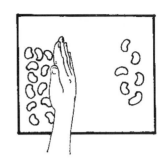

by Loree Saberin

Higher decade combinations

"Count up" from a specific number, e.g. 28 + 6.
They will add numbers together faster.

Materials

Per pair of students:
- 1 ziplock bag with objects inside
(such as beans) to be used as counters
- a counting mat (piece of paper or felt)
- overhead projector

Transitional activities

This activity will build on or increase skills in addition. They have already worked on small number addition. This will allow them to count higher numbers.

Classroom environment

This can be conducted in small groups of 5 - 6 students. Moderate noise level is to be expected.

Anticipatory set

With your partner, find 12 + 6. When you both know, raise your hands. (Give them a minute.) How many of you started at 1 when adding numbers together? How many of you ran out of fingers when adding? What answer did you get? We're going to play the Cover-Up game.

Purpose

I am going to show you a secret to adding faster.

INPUT	MODELING	CHECK FOR UNDERSTANDING
I'm going to give each of you some beans. Please leave the beans in the bag until I tell you what to do.	Hand out the beans.	Make sure they leave the beans in the bag.
Place the beans in a bunch at the upper right corner of your counting mat.	Put beans in upper corner of overhead screen. 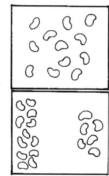	Make sure they have all placed their beans in the proper place.
Place "x" amount of beans in a pile on your counting mat. e.g. 12		Check to make sure that all students have "x" amount of beans on their mats.
Now place 6 more next to them on the mat.		Make sure they put out 6 more.
With your left hand cover up "x" amont of beans on your mat. e.g. 12.	Cover up beans. 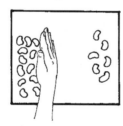	Make sure each child has covered the right amount of beans.
Ask the children how many beans are being covered with their hand.		Chorally: 12 All the students should have the same answer and together they should respond with "x".
Ask the children to start counting with the number being covered up and continue with the rest of the beans on the mat.	Start with the number covered and continue on.	Choral counting: 12.. 13, 14, 15, 16, 17, 18. Each student should finish on the same number. e.g. 18.
Now let's play in pairs. Pick a partner.		Make sure each child has a partner.
One person gets to decide how many to cover and the other decides how many to leave uncovered.		Check that each one has a job.

Number

INPUT	MODELING	CHECK FOR UNDERSTANDING
Pose problem for guided practice: Each person in pair will take a turn seeing what is the biggest she can fit under her hand. Then partner adds a number less than 10 and person counts up total.		
Now add.	Add the uncovered amount to the number covered.	The pairs should come up with the same number.

Guided practice

Have the whole class play the game a few more times. As a group, determine the number to be covered up.

Independent practice

Have the children play the game in pairs. Encourage them to share how they "count up" in each example, e.g. 18 + 3. Cover 18 and count "19, 20, 21"

Assessment

Check to make sure the children can add an equation like 13 + 4 orally by asking, "How much is 13 plus 4?" "How could you prove that?" Vary the numbers being used and the order in which the larger one is placed. See who starts counting from 1 rather than from the number under their hand. Children who are not firm in the higher numbers will falter as they approach the decade number e.g. 60 or 70.

Closure

Have the students explain what they learned about adding that will help them add better. Work with the children so that they realize they can start with the larger number no matter where it is in the equation.

Colliding Cubes

by Alan Hill

Sums of 11 through 20

Develop addition combinations using unifix cubes.

Materials

20 Unifix cubes per student or pair of students.
Combinations worksheet (see example).

Classroom environment

This lesson will involve students pairing up to work together. It will be noisy but at a controlled level.

Anticipatory set

Let's assume you have 3 Teddy bears and you have a friend who has 7 Teddy bears. If you take all of your Teddy bears to your friend's house, how many Teddy bears will there be together? Raise your hands, when you have an answer. (Ten.) Raise your hands if you agree! OK! What if you had 4 baseball cards and your mom bought you 6 more. How many would you have now? Raise your hands if you know. (Call on one student: Ten.) Does everyone agree? OK! Great! Ten is the correct answer.

Purpose

Today we are going to work with unifix cubes to practice counting so that we don't have to add on our fingers anymore. We'll find answers that use 10.

Number

INPUT	MODELING	CHECK FOR UNDERSTANDING
Let's count our Unifix cubes as I hand them out. You should each have 20.	Pass out Unifix cubes. Count 20 for yourself.	Listen for 20's. Correct any who don't have 20. Or count with them if they do have 20.
(Allow students time to play with them.)		
Now, I'd like each of you to count out 13 cubes and put them together.	Snap off 13 and hold them up.	Watch for 13 cubes.
I'll write the 13 here on the board like this.	Write "13."	Look for eyes on board.
Next let's take 3 cubes of to our original 13.	Write "10+3=13" on board. Hold up 10 + 3 cubes.	Watch for cubes for each student.
Now, how else can we make 13? Show your cubes when you know.	Show 13 cubes.	See that all students show cubes.
Call on several students.		
Thumbs up if you agree? Did anyone get a different answer? Emphasize how easy it is to "read" the number of cubes when you have 10 in a "stick" and the others seperate. Do a couple more.	Write equations on board. (e.g. 9 + 4 = 13)	See if all agree. If not call on those who raise hands and then count up with the class to get correct answer e.g. 9, 10, 11, 12, 13.
Repeat above with combinations for 14, 15, etc.		

Guided practice

Pair up students and pass out practice paper. Explain that one student makes the problem using the cubes. The other writes the equation as the first student calls it out. Student putting together the cubes should call out the equation to partner to write down. The student should work down the first column and then trade jobs so that the other student records in the second column. This is a cooperative exercise and not competitive. Students should be encouraged to help each other and work together. Each should count the number of ways, record it at the bottom of the column and then sign it. Have them do a couple with you.

Combinations

Independent practice

Let the students make the combinations with the cubes and record the equations on the practice paper. Circulate and assist as necessary. They may turn it over and make up their own problems if they finish early.

Closure

Ask student volunteers to write their equations for 13 on the overhead so they have all of them. Do the same for 14. Repeat for the patterns used on the practice paper. Ask the students what they learned.

Assessment

See if students can reverse process by making a combination into 10 + 3 e.g. 8 + 5: take 2 cubes from the 5 and put them with the 8. Do this for other sums. Can they do it concretely? Can they do it "in their heads?"

Missing Addends in the Little Red Hen

by Carrie Callett

Missing addends
Students will judge whether a missing addends equation matches at story. After listening carefully to a made-up problems related to the Little Red Hen a student volunteer will write the equaiton for the others to read for example: $4 + \square = 7$ or $9 - \square = 6$.

Materials

The Little Red Hen storybook
Chalk and Chalkboard

Anticipatory set

Students will sit down on carpet and I will start writing an equation on the board for them to answer (example above). This will get them focused on how to set up the equations when I read the word problems.

Input

"I'm going to read you a few story problems that have the same characters in them as *The Little Red Hen*. Who can tell me what the four characters are?" Write them on board. "O.K. Listen carefully and raise a quiet hand when you know how to set up the equations, or if you know the answer." I'll read different story problems. "Come up here (board) and write it out." "Is he/she right?" "Tell me with your silent signals." If the answer is wrong, "Can someone help fix it?" "How did you get your answer?" Have them explain it. "Did anyone else come up with another way to get the answer?" Ask several times. "Now, I'm going to write an equation on the board and I want a raise of hands if you know the answer." "I'm going to write some problems on the board and I want some volunteers to come up and fill in the answers." "Make sure you look at the signs to know if it's addition or subtraction." Ask how they got the answer - if wrong, someone can help him or her.

Model

I will show them how the equations tell the story. (3 + ☐ = 10. "There were 3 little chicks and some more came to equal 10.") (10 - ☐ = 6. "Ten little chicks and some went away leaving 6.") (They have done this type of work before, but some have added 3 + 10 and put 13 in the box. This practice will hopefully help them.)

Check for understanding

Have the students chorally count up (**3.** 4, 5, 6, 7, 8, 9, 10.) or count back (**10.** 9, 8, 7, 6) if they're not sure. Students will show with silent signals if the answers are right or wrong. If it is wrong, another classmate will have a chance to get the right answer (whole class will be watching). Having the students give reasons for their answers shows me their level of understanding.

Closure

Students chorally read the filled-in equations as teacher points to them e.g. "There were 3 little chicks and 7 came in so then there were 10." Let a student make up the sentence and then have the class say it chorally.

Assessment

Write 2 + ___ = 6 on the board. Using this as the title, have students write a sentence telling what happened to the chicks.

Tally Marks

by Patricia Marshall

Multiplication as repeated addition
Basic facts from 0x0 to 9x9

Students will create their own visuals for multiplication using tally marks. They will invent mental strategies to make counting high numbers easier.

Materials

Paper or card stock cut the size of cards
Rubber bands (to hold the tally cards in packs)
Two dice per student pair (or see p.88)
Use the sums of the dice to generate the numbers 0-9 by substituting the numbers 0, 1 and 2 (and later: 2, 3 and 4) for the sums 10, 11 and 12.

Classroom environment

Children are in pairs using 6-inch voices.

To begin

I will show the children two dice and tell them that we are going to play a game in pairs called Tally Marks. Then I will read a sum from the dice and demonstrate how to draw it using tally marks.

Give the children a purpose

You will learn shortcuts for counting high numbers.

Tell them/Show them/See if they are with you

Game: Tally Marks is a two-person game that works like this:
Toss the dice once and draw that many lines e.g. 4
Toss the dice a second time to determine the number of tally marks to make on each line e.g. 6
You will make six tally marks four times. Count up the tally marks and write the equation for the total number.

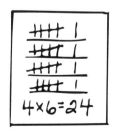

Whoever has the higher number circles the equation he has written on his tally card. After seven rounds each player adds up his tally marks and the one with the most wins.

Model this with several examples each time making a new tally card. Ask children after each roll of the dice, "How many lines do I draw?" "How many tally marks on each line?" "How many total?" "How did you get that?"

Let them play the game

Each player keeps his own stack of cards as they work in pairs. Observing and interviewing students as they work will give you some idea of the mathematical strategies they are using (e.g. simple counting, skip counting, repeated addition, commutativity, splitting the product into known parts, etc.) to shortcut the counting of the many tally marks.

Summarizing/Assessment

Share landmarks. "Who got 9x9?" "What did you do when you rolled a zero on the dice - how did you show it on the tally cards?" "Which ones did you like to do most?" e.g. the fives.
Journal entry: Prove that __ x __ = __   (e.g. 6x3=18).

Later lessons: Have the students explain their shortcuts in Student Presentation Lessons (see page 27): volunteers share a transparency of their work on the overhead projector e.g.:

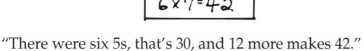

"There were six 5s, that's 30, and 12 more makes 42."

"I changed each 9 to 10 by taking 1s from the last line. (That gave me 60.) There were 3 left on the last line. So that makes 63.

As their stack of tally cards gets thicker they may also play Number Genius: students take turns in pairs showing a tally card with the equation hidden by their hand. The partner guesses how many lines and how many tallies on each line **without counting** (low-number genius). Then she guesses how many tallies total (high-number genius) and makes two piles of cards: easy and hard. The number geniuses analyze the hard ones together. No rushing!

Spatial Multiplication with Color Tiles

by Heidi Dettwiller and Caroline Miller

Multiplication as a rectangular array

Students will demonstrate understanding of multiplication in a visual and concrete form by showing the basic facts as rectangles on a grid. In pairs, students will use color tiles as manipulatives for creating various forms of rectangles to represent the basic multiplication facts.

Materials

For each pair of students:
25 color tiles
Grid sheet
Colored pencils
Blank Sheet

Introduction

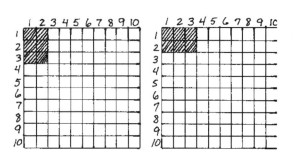

Students divide into groups of two. Begin discussion by showing the students how manipulatives can be used on the grid to create visual images of basic multiplication facts. Model for the students 2 x 3 and 3 x 2. Show students how to place three down and two across.

Ask students how many tiles are in the rectangle each time. Emphasize that this is the **same** rectangle rotated. It has the same identical shape except that the first is 2 x 3 and the second is 3 x 2. It has six tiles. Discuss how the same number might be represented **differently** (6 x 1). Show on tiles. Present the group with another fact and ask a volunteer to show how this fact could be represented with the tiles. (For example: 1 x 8, 2 x 4). Then give the group one or two more facts to practice with. Check for understanding as the student volunteers take turns with the manipulatives by asking the group if they agree or disagree (thumbs up/down).

Exploration

Ask the students how they could represent the numbers from 1 to 25 on the grid in as many **different** ways as possible. Discuss their answers, then model two examples 8: (2 x 4, 1 x 8) and 12: (3x4, 2x6, 1x12) using the color tiles. Show the students how to shade the rectangles on the grid worksheet to show the possible arrangements of tiles. Write the equation both horizontally and vertically near the grid representation. Give the students the grid and blank sheet. Ask them to use the manipulatives or grid paper to find all the different ways of representing each number. One student should use the manipulatives to find possible answers while the other records the possibilities on the blank sheet. Students can switch roles halfway through the practice exercises.

Summary

When all students have finished the practice sheet, ask students to share their answers with the group. Discuss the various ways of representing each number and each basic fact.

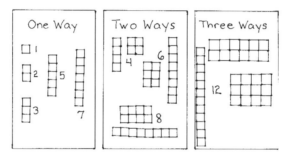

Ask students if anyone can show the group 0 x 3 on the grid, and have the class discuss why not. Ask students to explain what they noticed during exploring and discuss these with the group.

Extension

Have students do the same for the numbers from 26 – 81.

Assessment

Journal entry: Make up a problem that can be solved by the equation 5x4=20. Draw a grid to show the solution.

Source: "Productive Pieces: Exploring Multiplication on the Overhead" by Maureen Stuart and Barbara Bestgen. *Arithmetic Teacher*, January 1982.

LCM

by Dani Doiron

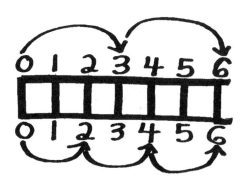

Identifying the least common multiple.

Materials

Paper, pencil and ditto

Transitional activities

This story problem develops the concept of LCMs through problem solving. The students have been using the problem solving strategies used in these problems.

Classroom environment

Groups of two will work together. Group discussion will be encouraged.

Anticipatory set

How long do you suppose it takes to make a pizza? Your answers may vary due to the ability levels of the pizza makers. This pizza problem will help us in reviewing the use of LCMs. Can anyone tell me what LCM stands for?

Purpose

This lesson is to help develop the idea of LCMs. As a result of this lesson you will be better able to determine the least common denominator of two or more fractions as well as to solve problems.

INPUT	MODELING	CHECK FOR UNDERSTANDING
Each person should have a pencil, paper, and question sheet in front of them.	Point to these objects as they are mentioned.	Look to see if they all have their supplies. question in front of them.
I would like for each of you to follow along with me as I read the problem. "At Eagles Nest Pizza..."		After allowing enough wait time, I will ask for a volunteer to summarize the question.
Does anyone have any suggestions on how they might set the problem up?	I will make eye contact and demonstrate that I am ready to begin recording any possible solutions.	Are students able to generate a response? Have volunteers share their ideas. Encourage them to draw pictures.
After student ideas have been shared, show them the number line, marking off 3s, 6s and 9s until they end together.	Label the number of pizzas made by each.	
I am going to split you into groups of two.	I will point to the individuals as I group them with their partners.	Check to see that students are attentive. Are they able to repeat to me who will be working with whom?
Please be sure that both you and your partner understand how the problem was solved.		I will watch and listen for team effort and understanding.
Remember that two questions need to be answered.	I will hold up two fingers.	Ask students how many questions need to be answered (choral response).
If you complete your work early see if you and your partner can make up another problem with the same answer as the first problem.		

Guided practice

I will give students a start by helping them identify helpful strategies (e.g. using the number line, drawings, tiles) for solving the problem. (There will be no guided practice.)

Independent practice

Students work in pairs to solve the two problems.

Closure

I will review the different ways that students came up with an answer. We will point out the easiest and most organized ways of getting an answer. Student volunteers will come up to the board to draw and explain their approach to the problem and the solution they found.

Assessment

I will have the students make up their own problem similar to the ones they solved in which the LCM is 36, for example. I will then have the students explain their problem to me.

Eagle's Nest and Carmichael School

At Eagles' Nest Pizza Parlor, Marla, John, and Mike work afternoons making pizza. If it takes Marla 3 minutes, John 6 minutes, and Mike 9 minutes to each make a pizza, how much time will pass before all three workers have a pizza ready at the same time?

How many pizzas will the three of them combined have done?

At Carmichael School's jog-a-thon Mark, Lynn, and Aimee are running to help earn money for camp. If it takes Mark 1.5 minutes, Lynn 2 minutes, and Aimee 3 minutes to each run a lap, how much time will pass before all three runners finish a lap at the same time?

How many laps will the three of them combined have run?

** If Mark has been running for 7.5 minutes, how many laps has Aimee completed in the same amount of time?

Number

Dividing a Number by Itself
by Paula Fallis

Division in two situations:
Partitioning (sharing) and Measurement

The children will create stories illustrating that a number divided by itself equals 1.

Materials

Beans and construction paper.

Anticipatory set

We have spent a great deal of time exploring multiplication through story-telling and because you did such a fabulous job creating those stories, I would like us to explore division with more of your creative stories! We will be searching and exploring what happens when a number is divided by itself.

Objective

Students will develop the concept of dividing a number by itself and discovering the answer is 1 using real world experiences. Students will have a deeper understanding of this concept through verbal discussion of their stories. Their stories should illustrate the two division situations, partioning, in which the objects are divided equally and the question is "How many are in each group?" and measurement, in which we know the number in each group and the question is "How many groups?"

Classroom environment

We will begin as a whole class and as we approach an understanding of the task they will work in pairs.

INPUT	MODELING	CHECK FOR UNDERSTANDING
Present story to students - "I have six jelly beans and I want to give them to six of my friends, how many jelly beans will each of my friends get?" Show me on your hands.	Write story on overhead. Write student response: 1	Students respond by holding up one finger.
Students create a story. Ask for ideas. We want our answer to be one.	Use overhead to write the story.	Student participation. Answer to story is one.
Ask for volunteer to explain reasoning.	Student draws a picture or explains using beans on overhead.	Observe students' faces.
Brainstorm for ideas. Beans can represent anything.	Draw pictures based on their ideas.	Students use imagination in brainstorming.

Guided practice

When you and your partner receive your beans and construction paper you will create one more story together. "Do we have enough beans?" "What shall our beans represent?" "How many shall we use?" Write a story that uses division. You may arrange your beans the way you would like. Be creative and remember that the answer to your division problem should be 1. One partner writes the story which will end in a question. The other partner draws the picture.

Independent practice/Assessment

Students write their own story and represent it with a drawing. Does it require division for a solution and is the solution 1?

Closure

Students share stories. "What discoveries did you make while creating your stories?" "Did anyone come up with an answer other than one?" Discuss the stories. Show how to write division $6\overline{)6}$ or $6 \div 6 = 1$ two ways and read a story pointing to each number as you read.

Word Problems in Multiplication and Division

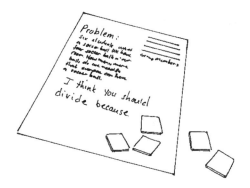

by Suzanne Blakeney

Choosing the right operation
Solving word problems

Students will work in pairs and use problem-solving strategies (acting it out, drawing a picture or using manipulatives) to choose the right operation to solve word problems. This lesson is used in conjunction with a textbook page of mixed word problems (multiplication and division).

Materials

Pencil, paper, textbook, bell. Also, have these materials available for students to gather during **exploring**: writing paper, grid paper, color tiles or beans.

Transitional activities

I'll relate this lesson to the multiplication and division they have been studying.

Classroom environment

Quiet talking during groupwork.

Anticipatory set

I'll tell the students that I am going to use them to help solve some problems by acting them out.

Purpose

We run into problems that can be solved by math all of the time and we need to know how to begin to figure them out.

Introduction

- I'll start by instructing students to clear off their desks.
- I'll explain that multiplication is when you have some groups, all of the same size, and you want to know how many there are total. Division is the opposite of multiplication. In division, you know how many there are total and you want to know how many groups there are or how many in each group.
- I'll model three strategies for figuring out problems:

Strategy 1 - **Acting it out**: (Call ten students to the front of the room to act this out.) Two of the students are famous people. (Name them.) The other eight students wish to shake hands with the famous people. How many handshakes will the famous people make? Let's count the handshakes. (Count chorally.)

O.K. so how many handshakes total? (Chorally: 16) We had two numbers, 8 and 2, in this problem. What did we do with them to solve our problem? Get several explanations e.g. "We added 8 and 8." "We multiplied 2 x 8." "We added." "We multiplied." Record e.g. 8 + 8 = 16 or 2 x 8 = 16.

Seat the students for the second strategy.

Strategy 2 - **Drawing a picture**: The whole class (e.g. 32) is going on a field trip. Only three students can go in each car. How many cars do we need? (Record their initial guesses. Then on the front board draw "stick people" representing the class and draw cars around every three.) How many cars? (11) We had two numbers, 32 and 3 in this problem. What did we do with them to find the answer of 11? Record their answers e.g. 32 divided by 3 = 11 or 3 x ∆ = 32. Discuss why a fraction was not used in the answer.

Strategy 3 - **Using rectangles on grid paper** or **beans in groups**: Our class is going to get into 8 equal groups. How many students are in each group? If we used this grid paper and made 8 rows, how many squares would be in each row? (Try 3 in each row first and ask if this is right. Let students direct that 4 squares go in each row.)

"You are going to work in groups of four to solve a problem in your textbook. You and your partners will solve the problem using one of these problem-solving strategies and record your work on a sheet of paper." Model the format of the paper on the overhead. "When you are done, each group will read their problem to the class and show how to solve it using one of the strategies. Your purpose is to help the class to understand your problem." Direct them to a page of mixed word problems and distribute problems among the groups. "When you hear the bell, that means you should stop what you're doing and listen."

Exploring

Students work in pairs to solve one problem.

Summarizing

Ring bell. Turn on the overhead and tell the students we are ready to start. Ask for a volunteer to go first to read their problem. "What did you do, multiply or divide?" "What strategy did you use to solve the problem?" "Will you show us?" "What answer did you get?" "Are there any questions from the class?" Let every pair that worked on the same problem share and compare what they did. **Ring the bell when discussion gets too noisy.**

Assessment

After students have used the strategies over several days, ask them to write a paragraph discussing what strategies they have actually used and how they feel about solving problems.

Explaining + and - Situations with Base-10 Materials

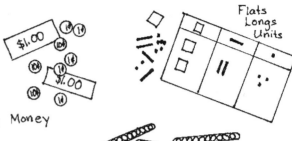

by Pat Marshall

Two-digit addition and subtraction
Children explain their processes using base-10 materials

Materials

For Students
Base-10 mat for each pair (appendix)
Base-10 materials for each pair (appendix)
(e.g. 2 flats, 20 longs and 20 units)
Worksheet (enclosed) and pencil

For Teacher
Transparency of base-10 mat and felt pen
Base-10 materials
Bell

Transitional Activities

Students have been introduced to base-10 materials and have had some free exploration with them beforehand.

Classroom Environment

The introducing is teacher-directed. Students are listening to the teacher, watching the overhead and answering chorally. They also pair-share and explain to the whole group. The summarizing is student-directed , the explanations being done solely by students with the teacher acting as moderator. (It will probably take several days to accomplish everything suggested in this plan.)

Number

Anticipatory Set

"I am going to show you a number on the overhead. See if you can read it." Flash the overhead projector on and off several times, each time revealing a different number with the manipulatives. "What number was that?"

Purpose

We are going to solve some real problems that we have had in our class using our base-10 materials.

Introducing

INPUT	MODEL	CHECK FOR UNDERSTANDING
Problem 1 The entire fifth-grade brought snacks. How many snacks?		Chorally: "32."
Students from another class also brought snacks. How many?		Chorally: "16."
How many snacks total?		Chorally: "48."
Problem 2 We got this many announcements to hand out. How many are there?		Chorally: "65."
How many students are in our class?		Chorally: "33."
So if everyone distributes one, how many more need to be distributed?	Remove one piece at a time.	Chorally: "55, 45, 35, 34, 33, 32!" or Chorally: "64, 63, 62, 52, 42, 32!"

INPUT	MODEL	CHECK FOR UNDERSTANDING
Problem 3 Eight carloads of students went camping. How many people?	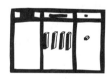	Chorally: " 41"
Next day it was very cold and some went home. Share your answer to this question with your partner: "How many left the next day if there were this many remaining at the campsite?" When you both agree, raise your hands.	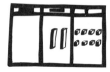	Pair share. Volunteers. Choose several to give a different explanation of how they figured out how many left.

In later lessons, use problems that involve zeros.

There are 205 students at Creekside School but 26 are absent. How many students are at school? Here are my 205. What will we need to do to solve this problem? How many agree?		Volunteer: "Subtract" See whose thumb goes up immediately.
Yes, and looking at this problem we can see that subtracting 26 will take some exchanging frst because we don't have enough units. Take 30 seconds and discuss with your partner how you would subtract 26 from this number.		Students pair-share their ideas. Then show.

Discuss and model the regrouping before presenting this problem to be solved in pairs using manipulatives:

> There were 28 cars in the parking lot and 17 more drove up. How many are there now?

The signal to stop exploring and listen to the teacher is a bell.

Exploring

The children work in pairs to solve the problem with the idea that they will present their solutions to the whole class.

Summarizing/Assessment

Have students write a story involving addition or subtraction (their choice) and then explain how to solve it with or without manipulatives.

_____
_____

Our Problem

_____
_____
_____
_____
_____

To solve this problem, you have to _____ .

We did it this way with our base-10 materials:

[ ]

Picture of Our Solution

_ _ _
Answer

Using the Hundreds Chart to Add and Subtract

by Pat Marshall

Develop mental strategies for two digit addition/ subtraction with and without regrouping

This lesson is mainly guided practice with some pair-sharing. It is done orally and visually, each child using his own hundreds chart to add and subtract such problems as 35 + 29 = 64 and 45 - 18 = 27.

Materials

Hundreds Chart for each student (appendix)
Hundreds Chart transparancy for teacher
Overhead projector

Transitional activities

Students have made their own hundreds charts and have noticed lots of patterns in them e.g. that all the ones digits in each column are the same; adding 10 to a number takes them to the number beneath it. They know their basic addition/subtraction facts.

Classroom environment

This lesson is entirely teacher-directed. Students are listening and watching as the teacher demonstrates using the hundreds chart on the overhead. During the recitation, students either call out answers, point to them on their hundreds chart, or explain their methods in pairs (pair share) or to the class.

Anticipatory set

Flick on overhead projector showing hundreds chart. "Remember making these hundreds charts? What patterns do you see?"

Purpose

Today you are going to use the hundreds chart to learn to add numbers like 35+21.

INPUT	MODEL	CHECK FOR UNDERSTANDING
First, I will find 35.	Put finger on 35.	
Then, I will add the easy part of 21 first. Which is easy for you, adding 20 to 35 or adding 1 to 35?	Show adding 20 first, then 1. 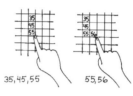 Then show adding 1 first, then 20. 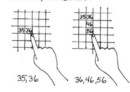	Have a show of hands to vote for either 20 or 1.
How about 56 - 19? (Ask how rounding one of the numbers makes subtracting easier.) Will we move up or down to subtract? (Model other examples as necessary.)	Write 56 - 19. Round 19 to 20 and show on chart subtracting 20 and adding 1. 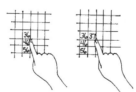	Volunteers tell teacher which number to round and direct her where to move her finger on chart.
Do you think you could do this on your own hundreds chart?		Call outs: Yes! (Monitors distribute charts.)
Get with a partner. You will be showing your partner how you personally solve the next problem. Raise you hand if you still need a partner.		Students choose partner. Pair up any without partner.

Guided practice

In the remainder of the guided practice, give the students two or three problems of addition or subtraction, each time letting them show their partner how to solve the problem. When both are satisfied that they have the correct solution, they raise their hands. Show solutions on the overhead and move to the next problem.

Closure

Raise your hand if you think you got a little better at adding or subtracting today.

Assessment

Observe how they use the hundreds chart. Do they add and subtract multiples of ten easily? Do they show flexibility in working with either the tens or ones first? Do they make good decisions in rounding e,g, rounding 19 to 20?

Showing Partial Products Using Rectangles

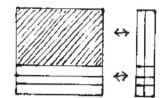

by Pat Marshall

Constructing products of 2-digit numbers

Students will use base-10 materials to make rectangles that show products and will analyze them to show the partial products.

This lesson uses guided practice with some pair-sharing. It is done orally and concretely. Each student will use base-10 materials to construct rectangles that show a specific product, e.g. 12 x 13 = 156 and then describe the two smaller rectangles that make it up.

Materials

For Students
Base-10 materials for each pair
(e.g. 1 flat, 11 longs and 30 units)
Paper and pencil

For Teacher
Clear transparency and felt pen
Bell

Transitional activities

Students have used base-10 materials before for addition and subtraction and they know the value of each piece. They have made rectangles to show smaller products e.g. 2 x 3 = 6.

Classroom environment

This lesson is teacher-directed. Students are listening and watching as the teacher demonstrates using the base-10 materials on the overhead. During the recitation, students either call out answers, show them on their hands, or explain them to the class or to each other in pairs (pair share).

Number 339

Anticipatory set

Show a 3 x 2 rectangle on the overhead projector. "What are the dimensions of this rectangle?"

Purpose

Today you are going to understand how to multiply two-digit numbers like **1 2**
                                                                      x 1 3

INPUT	MODEL	CHECK FOR UNDERSTANDING
Review products of small numbers: e.g. 3 x 2 and 2 x 3.	Show on overhead.	Students call out products e.g. 6.
Now show me what would 3 x 10 look like? 10 x 3?	Show on overhead.	Class holds up 3 longs.
I am going to make a rectangle that is 13 x 12. Discuss with your partner for 30 seconds what pieces I might use.		Students pair share.
What do you think?		Volunteers share their ideas with the class.
Let's use this one. How many units are in this rectangle? Everyone see that?	Show on overhead.	Volunteer: 156. Thumbs up/down.

INPUT	MODEL	CHECK FOR UNDERSTANDING
Where do you see the 13, along the bottom or along the side of the rectangle?		Call outs: Bottom.
I'm going to split the 13 into 2 parts, the 10 and the 3. Now you make this rectangle and then share with your partner: How much is in each smaller rectangle?		When partners are finished exploring the rectangle, they will share with the whole group what they found.
O.K. Now what about the small rectangles in this one.	Show a 12 x 13 rectangle. 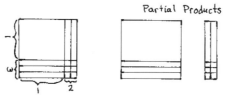	Let them discuss as before.

Independent practice/Assessment

Make a rectangle and find its product and then divide it in 2 parts as we did before. See if the student can show the partial products using the materials. The bell ringing signals: quiet and listen.

Closure

Choose one of the student-constructed rectangles e.g.

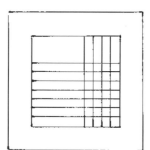

```
          1 7
        x 1 4
```
and show how the two smaller rectangles have these partial products:
```
          6 8
        1 7 0
```
And this is the product for the whole rectangle: 2 3 8

Dividing Money from a Cash Drawer

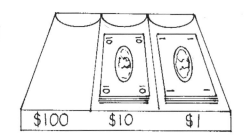

Short division with 1-digit divisors and 2-digit dividends

Students will orally describe how to divide money into equal amounts using ten- and one-dollar bills in a cash drawer.

Materials

Cash drawer (made from cereal box) for each pair of students
Fake money (see pp. 296-299)
9 ten-dollar bills and 20 one-dollar bills for each pair of students
(Note: Make each denomination a different color)
A blank piece of paper for recording
Transparencies of cash drawer and money for the teacher

Transitional activities

(How is this related to other lessons?) Students have already used the divided mat with some base 10 materials (e.g., money, bean sticks, FLU materials) and know how to make numbers on the mat. The cash drawer is a variation of the 3-place mat or FLU board. They also know how to exchange or regroup 10 ones for a ten-dollar bill and 10 ten-dollar bills for a hundred-dollar bill. (See **Explaining + and − Situations with Base-10 Materials**.)

Classroom environment

Quiet during introducing and noisy (12-inch voices) during exploring. There should be a mixture of noisy and quiet during summarizing since the class should listen as each student speaks but will be more animated as they agree or disagree after each procedure is explained.

Math Plans

Anticipatory set

I'm sure you've all noticed that the cash drawer in a grocery store cash register is organized to keep each denomination of money separated. We're going to use a simplified cash draw – no $5 or $50 bills – to divide up some money in real life situations. (Show your cash drawer transparency on the overhead projector.)

INPUT	MODEL	CHECK FOR UNDERSTANDING
Problem 1 There are three days left until payday and John has $36 left. How much is available for each day if each day he spends equally?		
The problem can be written mathematically like this or this.		
We will be using our money and cash drawer to solve this problem. I'll show you my way a bit later. You show me your way using your cash drawer.		
First I'll put $36 in the cash drawer.		
The problem is to put 36 into groups of 3. I'll start with the $10 bills and split them in 3, one for each day. I'll put them above my cash drawer. How many tens for each day?		Chorally "1"
So far we've split up only the tens in our cash drawer. We still have to split up the ones. Let's do that. We can put $1 for each day. But we still have $3 left so let's put another dollar for each day. So now how many $1 bills for each day?		Chorally "2"

Number

INPUT	MODEL	CHECK FOR UNDERSTANDING
Now I want you to "pair-share" what you think with your partner about the answers to these questions: How much can John spend each day? Where do you see the answer?		Students discuss for about 1 minute how much can be spent in a day ($12) and can see this by horizontally reading the money on the overhead.
Let's see what you think. Aesha, how much can John spend and show me up here where you see that.	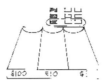	Student answers e.g., "12" and points to this or one of the other two rows.
Raise your hand if you can see the $12 Aesha showed.		Scan the class to see whose hand is not up to determine if one or two more simple problems are needed, e.g., 2 days with $26. 4 days with $44 left.
That answer is written like this. This means that there was 1 ten dollar bill and 2 one dollar bills for each day.	$$3\overline{)36}^{\,12}$$	
Problem 2 This time you will use your cash drawer to do the problem with me. So open your bag of money and put $42 on your cash drawer.		
Your cash drawer looks like this.		Check to see that everyone can do this easily. It should look like this.
You and your best friend want to split this evenly. First the ten dollar bills then the one dollar bills like this. Go ahead and split up the money if you have a way to do it.	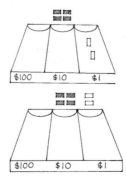	They may get noisy here especially if a lot of them know how to do it.
So how much money does each friend get? Tell me what to do.		Chorally "21"

INPUT	MODEL	CHECK FOR UNERSTANDING

We'll write the answer like this to show there are two ten-dollar bills and one one-dollar bill.

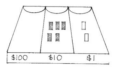

Okay, now comes the harder one. Suppose instead of $42 they had $52 to share. Show me $52 in your drawer. Here's what it looks like.

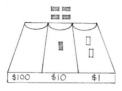

They each add a ten dollar bill.

Let's divide up the ten bills.

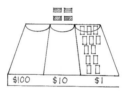

We have a little problem here though. There's a ten dollar bill left and we want to split it. There are several ways we could do this and I want you to be thinking of how **you** would do this. I'm going to "exchange" the ten dollar bill for 10 ones in the cash drawer with the other one dollar bills.

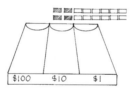

I divide the dollar bills in half and it looks like this.

So now, how many of you think you know how much money each person got?

Children raise hands. Keep an eye on any who don't so that during exploring to make sure they are involved in trying the problem.

Everyone: How many dollars?

Chorally: "26."

Problem 3
(3 names of students in the class) wanted to share a $99 prize from a school raffle because they pitched in to buy the ticket that won. How much will each one get?

3)99

Number

INPUT	MODEL	CHECK FOR UNDERSTANDING
You will have 5 minutes to work on this problem. When you hear this sound (e.g., rhythmic clapping) you should stop talking and look at me.	Model the jobs of "doer" and "recorder" if they haven't done this: Doer touches the money to solve the problem. Recorder writes the problem and solution.	

Exploring

Look for these things in student groups:

Everyone participates in each group especially the quiet ones and the non-English speaking or limited English-speaking. Keep an eye out for students who didn't really seem to follow your introduction. Are they active in listening and manipulating in their group?

Look for this in the mathematical work with manipulatives: A variety of ways to explain how to divide $99 and get $33 each.

Student 1: Just split them up into three piles. Each pile has $33.

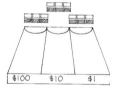

Teacher: How many did it this way? (Hands)

So, how many stacks of three tens? (Chorally: 3). And the same for the ones? (Chorally: yes). Any questions for (student 1)?

Student 2: I stacked the tens in 3s like this and I did the same thing to the ones. So I got $33.
$33 for the first person.
$33 for the second person.
$33 for the third person.

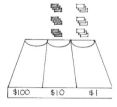

Teacher: How many did it this way? (Hands)

Any questions?

Student 3: I just looked at it and I knew that splitting it into 3 parts, each part would be 33.

Teacher: Could you <u>prove</u> it? How? Student attempts to prove.

Student 4: I knew that 3 x 33 is 99 so I just put 33 like this.

Yes.

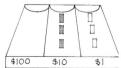

Teacher: You showed the answer <u>inside</u> the cash box instead of above it?

Follow-Up

In later exploring lessons, use problems with 2- and 3-digit dividends with and without regrouping (e.g. $4\overline{)412}$, $3\overline{)351}$) and with zeros in the dividend or quotient (e.g. $2\overline{)204}$, $5\overline{)105}$, $3\overline{)327}$). Test out the problems beforehand to be sure the students have enough manipulatives to do the regrouping!

$6\overline{)72} \quad 3\overline{)75} \quad 8\overline{)96}$

$4\overline{)56} \quad 3\overline{)39} \quad 5\overline{)65}$

Assessment

Have students use their manipulatives to make up a story about dividing and solve it. Then have them write out their story and its solution.

Hundreds of Dollars in the Cash Drawer

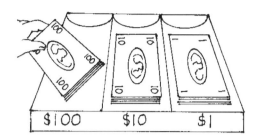

Short division with 1-digit divisors and 3-digit dividends

Students will orally describe how to divide hundreds of dollars into equal amounts using hundred-, ten- and one-dollar bills in a cash drawer. Students will orally give reasons as to where the quotients should be written. Students will record the quotient including fraction remainders.

Materials

Cash drawer (see previous lesson) for each pair of students
Fake money (see previous lesson) 9 hundred-dollar bills, 20 ten-dollar bills, and 20 one-dollar bills for each pair of children
Blank sheet of paper for writing problems and quotients
Transparencies of cash drawer and money for teacher

Transitional activities

(How is this related to other lessons?) Students are already comfortable with dividing two-digit numbers using the cash drawer. (See **Dividing Money From a Cash Drawer**)

Classroom environment

Generally quiet during guided practice except when students pair-share. Noisy during independent practice. Animated during closure.

Number

Guided practice

INPUT	MODEL	CHECK FOR UNDERSTANDING
Problem 1 A $572 door prize was divided evenly among the students in grades 4, 5, 6, and 7. If each student gets $4, how many will receive a prize? What do we have to do? Who can say it a different way? I'll put $572 in my cash drawer.	Write: 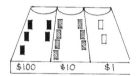	Volunteer 1: We need to divide 572 by 4. Volunteer 2: We need to find out how how many $4 are in $572.
Now we're going to divide the cash in each section of the cash drawer into stacks of how much?		Chorally: $4
And we want to find out how many stacks, right? That tells us how many children will get $4.		Silent assent.
Based on what we did before with smaller numbers, what do you think we'll do first? (Choose several to answer.)		Individual volunteer: I think we divide up the $100 bills into stacks of 4.
I'm going to do that.	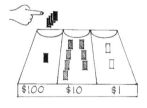	
Pair-share this question: How many kids are getting $4 so far? Point to the 400.		Students discuss with a partner how many students will get $4 from $400.
What did you decide? Thumbs up if you agree. I'm going to write 100 by putting the 1 here.	4)572 with 1 above	"100 students" (Observe to see whose thumb doesn't go up.) Have several volunteers explain their thinking.
The answer will be "1 hundred something" so let's figure out the "something." Trade in the remaining hundred dollar bill for how much, class?		Chorally: "10 ten dollar bills"

INPUT	MODEL	CHECK FOR UNDERSTANDING
And where do we put the ten dollar bills?	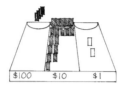	Chorally: "In the ten dollar section."

As before, we divide up all the tens into stacks of 4 and then we do the same with the ones. I'm going to stop here and have you form the number 572 in your cash drawer with your partner and solve the problem: How many children will get $4? When you have an answer, both of you raise your hands and I'll be by to see what you did.

Independent practice

Students work in pairs to finish solving the problem. When both of them can explain it, they may call the teacher over to show what they found.

Extension

Those who finish very early may draw their results. OR They may, at the discretion of the teacher, help another pair to work out the problem but they must use questions only. They may not tell them anything.

Closure

Students share their answers and how they did the dividing on the overhead projector.

Assessment

Observe children solve a problem using manipulatives during independent practice.

Future lessons

To explain long division, tell the students that this time you are also going to record **how much is left** in the cash drawer each time you divide. Ask these three questions as you finish with each section of the cash drawer:

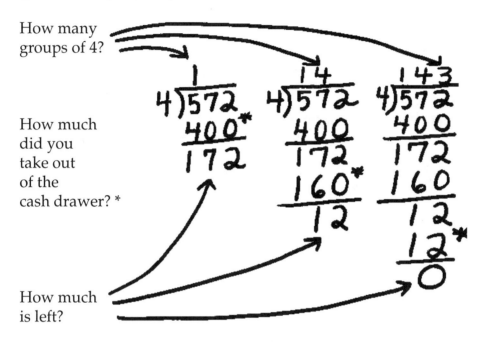

How many groups of 4?

How much did you take out of the cash drawer? *

How much is left?

Once they understand, using their manipulatives, what all the symbolic "stuff" hanging out of the bottom means, they are ready to see a shortened form, e.g.

Division with 2-digit divisors can be shown the same way. The difference is that more mathematical power is needed in order to think in chunks of 40s, 60s, or 80s rather than 4s, 6s, or 8s, and you will need a lot more $10s and $1s.

Fraction Freddy

by Karen Wong

Naming fractional parts of group of objects.

Materials

Stuffed teddy bear, flannel board with different
types of berries and some honey, packages of gummy bears, Freddy's Gummies (worksheets), pencils.

Transitional activities

(How is this related to other lessons?) This will be a review of fractions. They have already worked on a fraction packet.

Classroom environment

The noise level will be a little bit loud at times.

Anticipatory set

I will bring in Freddy (a stuffed bear) and introduce him to the class. I will tell a story about Freddy.

Purpose

Learning fractions will benefit them because they occur in everyday life. They need to learn fractions for cooking or baking, to handle money (what is a quarter of a dollar), and also to do business when they begin to work.

Input

Today we're going to review some more fractions. I brought a friend with me to help. His name is Freddy. He's really shy, so you have to be quiet and not scare him. He's going to help us learn fractions. (Then I will bring Freddy out and begin a story.) Freddy woke up early one morning before anyone in his house. He was really hungry. He was so hungry that his stomach was growling. He looked around in the cave and could only find 4 jars of honey (Put the honey on the flannel board.) He decided to eat 1/2 of the honey and save the rest. How many jars did he eat?" (Have a volunteer show them by taking the jars of honey off the flannel board. Ask if the children agree or disagree). Continue the story with Freddy going out and looking for some more food in the woods. On the way he finds some strawberries (8 or so) and then he decides to eat some (2/8 or 1/4). Have a child go up to the flannel board and take that number down. Then, I will ask if they agree or disagree. Then Freddy sticks the rest of the berries in his pocket but he doesn't know that he has a hole in his pocket. So, some of the berries drop out (let's say 2/6 or 1/3). I will go as long as I need to. When I feel that they are catching on, I will end the story.

After I finish the story, I will give them instructions about counting and recording Freddy's gummies. I will have an overhead projector for my worksheet, and show them step by step what to do.

Modeling

I will cut open the gummy bears package, count and sort the gummy bears and write the number on the worksheet on the overhead. Then I will count out the red ones and show them how to make the fraction. I will count the yellow ones and see if they can make the fraction and show me where to write it.

Check for understanding

I will have individuals come up to the flannel board and have them take down a given fraction of the berries. I will ask the class to show me with their hand signals if they agree or disagree. I will also ask them what they are supposed to do first when they receive their gummy bears. What is the next thing they are going to do?

Independent practice/Assessment

Student helpers distribute the worksheets and gummy bears. Students will count the total number, count the number of each color and write the number on the line. After they finish each color, they will write the fraction of each color on the next line. After they have finished, they are to raise their hand so I can check it. Ask them to give the fraction name of a given color, e.g. what part of the set is green?

Closure

I will review with them their results.

Freddy's Gummies

1. Sort your colors here.

Red	Yellow	Clear	Green	Orange

2. Count the total number of gummy bears. How many are there?

3. Write the number of each color and the fraction.

Red   _____   _____

Yellow   _____   _____

Clear   _____   _____

Green   _____   _____

Orange   _____   _____

_____
Name

The Doorbell Mystery

by Arlis Groves

Recognize a fraction as a portion of a whole or as a part, and of a group of things. As an extension, students will name the fraction and write it.

Materials

The Doorbell Rang by Pat Hutchins
Cookies - approximately 40
Markers - 8
Paper cookies - 40
Scissors - 8 pair
Glue - 8 bottles or sticks
Fraction Worksheet - 34

Transitional activity

Review model of circle divided into equal parts. Review what the students already have learned about naming and writing fractions. Call on students to draw a fraction and then ask another student to name it and come up and write it.

Classroom environment

Noise level must stay at a low "buzz" since this is a "pod" style classroom and sound carries to other classes. Arrange groups of four ahead of time.

Anticipatory set

Read *The Doorbell Rang*, by Pat Hutchins. The students should be able to catch on to the pattern and fill in words if I pause during the reading.

Purpose

Explain that there are many times, just like the story, when we must equally divide a whole e.g. unexpected guests for dessert. (Pie!)

INPUT	MODELING	CHECK FOR UNDERSTANDING
Have students form one large circle for the story.		Students are quiet and ready.
"We are involved in... *The Doorbell Mystery!* Before I tell you about the mystery we will solve, we've got to go through detective training! Watch carefully so you will have the tools you need to get to the bottom of this."		
On board, make a giant cookie and divide various ways. Who can name the part of the whole that I have colored in?"	Use marker to color. 	Call on a quiet, raised hand.
"Who would like to draw a cookie fraction on the board and then call on a classmate to name it?" Choose a volunteer.	The volunteer draws another cookie, divides and colors it. Then calls on a friend.	
"Does everyone agree? Show me with quiet hand signals. Does anyone have a different answer?"		Thumbs up/down. Call on a quiet, raised hand. Student who has been chosen answers.
Discuss this.		
Student calls on another student to create cookie fractions and so on		

Number

Guided practice

"Move into small groups of 4 and 5. All jobs should remain the same."

        1. Leader
        2. Recorder
        3. Reporter
        4. Observer

Explain that the Leader will draw divisions and color in a portion. Groups discuss how to write it and Recorder writes the answer. Reporter will tell about it. Observer encourages and keep others on task.

Hand out scratch paper. Hand out one marker per group.

Model the jobs with four students by using the example of dividing two cookies fairly among four people. Model again if necessary.

Independent practice

"Put each group member's name in a box on the chart ditto. Now that you are prepared for the challenge, your mission is to figure out how to divide these five cookies 4 ways equally!"

Hand out a baggie with 5 paper cookies inside to each grounp. Hand out chart ditto, scissors and glue.

"Put each person's share in his/her box on the chart ditto. When you are satisfied that you have solved the mystery, glue the pieces down."

Circulate. Check each group's work as it is in progress. Is each member doing his/her job?

Closure

"What did you find? Let's hear from each group. Reporter from group one...?" Discussion follows. Now explain to groups that the real cookies are going to be passed out. The same divisions should be made with these, as were made with the paper cookies. This time everyone gets to eat their share! Make sure the cookies are soft enough to cut.

Assessment

Use paper plates as pizzas. Give each student five paper plates and whisper a "secret mission": e.g. to divide their plates into thirds. They may fold them. Ask how many pizzas are in each "third."

The Doorbell Mystery

Show how your group divided your 5 cookies equally.

Tell how you did it.
_____
_____
_____
_____
_____

Names
_____
_____
_____
_____

Equivalent Fractions

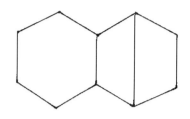

by Karen Wong

Making equivalent fractions for 1/2 using pattern blocks.
Comparing the equivalencies symbolically. e.g. 1/2 = 2/4 = 3/6

Materials

1. Transparency and overhead projector
2. Pattern blocks
3. Ditto - copy for each child
4. 4" x 18" colored strips of paper for Fraction Strips
5. Dice marked 1/2, 1/4, 1/8, 1/8, 1/16, 1/16

Transitional activities

The class has done similar work with pattern blocks for "1/4."

Classroom environment

This will be whole group instruction. They will be answering questions and constructing with pattern blocks. The noise level will be medium.

Anticipatory set

Today we're going to work with more pattern blocks! (I will put some on the projector and ask what the shapes are.)

Purpose

To expose the children to another model of equivalent fractions.

Input/model/check for understanding

Who can remember how many blue parallelograms it takes to make one yellow hexagon? (3) I will also ask them how many triangles it takes and how many hexagons it takes to make a yellow hexagon.

I will put them on the overhead and ask if they are all the same and equal. (Chorally: "No!") I will then have the "table captain" come up to get one for each member for the table group. I will show them how to do the first problem on the sheet and watch them step by step as they do it with me.

I will be asking questions and watching them place the manipulatives.

I will let them do the next problem and then have them tell me the answer. "1/2 = 3/6" "What will you use for the '1/2s'?" Chorally "Hexagons." O.K. Do it. What will you use for "1/6s? Show me the piece in your hands. (Shows parallelogram.) "O.K. Do it. Then be ready to tell me how many parallelograms make 1/2 and how your paper shows it." ('3'. Call on several volunteers to explain.)

I will show the manipulatives on the overhead after they have done theirs.

I will write some fractional equivalents on the overhead and ask what they notice about the fractions. Then show them that you must multiply both the numerator and the denominator by the same number to get equivalent fractions.
$$1/2 = 2/4 = 3/6 = 6/12$$

Closure

I will ask them to tell me what they must do in order to write an equivalent fractions for 1/2. They must multiply the numerator and the denominator by the same number.

Follow-up

Make a fraction kit using fraction strips (see Fraction Strips) to continue naming and comparing fractions.

Fraction Strips

Children play in pairs, each with his/her own game board: the "whole" strip is the game board. The goal is to be the first to cover the game board with fraction strips. No gaps and no overlapping strips are allowed.
1. Students roll cube and read number that comes up on top.
2. Puts that strip on the game board.
3. Each partner takes turns doing #1 and #2 until one of them "covers up" the entire game board. This person is declared the winner.

4. To make the game more interesting you can add the rule that you must get the exact fraction at the end in order to win. e.g. You need 1/16 to win but you roll 1/2. You must skip your turn and try again.

Assessment

Have student use pattern blocks to prove to you that 1/2 = 3/6.

Pattern Blocks Halves

Fractions that are the same part of a whole are called *equivalent* fractions.

Cover the figures below with pattern blocks. Show that the fractions are equivalent.

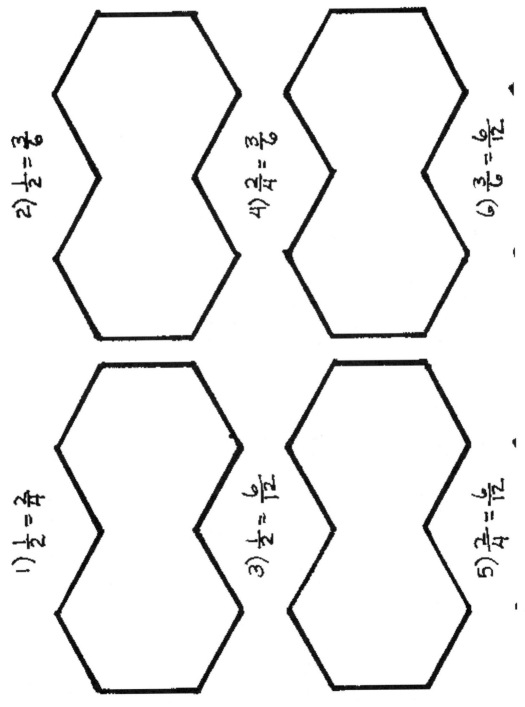

1) $\frac{1}{2} = \frac{2}{4}$

2) $\frac{1}{2} = \frac{3}{6}$

3) $\frac{1}{2} = \frac{6}{12}$

4) $\frac{2}{4} = \frac{3}{6}$

5) $\frac{2}{4} = \frac{6}{12}$

6) $\frac{3}{6} = \frac{6}{12}$

Geoboard Fractions

by Carolyn Hoffman

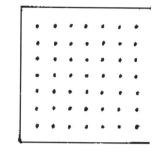

Naming and comparing equivalent fraction

Students will give their own definition of the term "equivalent fraction" and will show some equivalent fractions on geoboards.

Materials

Geoboard or geoboard dot paper (appendix)
Worksheet

Anticipatory set

I want all of you to picture a pie in your head. Choose your favorite. It could even be pizza pie. Now imagine cutting the pie into 8 equal pieces and picture 4/8 of it. Now picture 1/2 in your head. Which was easier to picture? These are actually the same amount of pie, one is just easier to picture than another. Since we have been learning about fractions and how to determine which fraction is larger than the other, it is now time for all of you to learn about equivalent fractions.

Purpose

By learning about equivalent fractions, you will be able to write very complicated looking numbers in a neater, simpler way.

INPUT	MODELING	CHECK FOR UNDERSTANDING
I want each group to pick a person to come pick up a geoboard for each pair of students.	Point to the cabinet where the geoboards are stored.	Students pick someone at the table to pick up geoboards.

Number   365

INPUT	MODELING	CHECK FOR UNDERSTANDING
We need to put the rubber band around the outside of the geoboard, the side which has a square perimeter.		Students should be placing a rubberband around the outside perimeter of the geoboard.
Now I want each of you to divide the geoboards first into halves and then into quarters using the rubberbands at your tables.		Students divide their geoboards into halves first and then into quarters.
Cover with your hand one half of the geoboard.		Students cover one half of the board with one hand.
Now I want each of you to tell your partner how many quarters or 1/4's you see.		Students quietly talk with one another.
Show me on your hand how many.		Look around to check students are holding up two fingers.
You are absolutely correct, there are two 1/4's in 1/2. Give yourself a silent pat on the back.		Students should give themselves a silent pat on the back and then look at the teacher.
Now with your partner decide what you need to do to divide your geoboard so that it is divided into six equal parts. (Have volunteers explain what they did.)		Students divide the board into 1/6's. e.g.

INPUT	MODELING	CHECK FOR UNDERSTANDING

INPUT

Good. Cover 1/2 of your board as we did previously so that only 1/2 of the board is showing.

Discuss with your partner how many 1/6's are showing in 1/2 of the board and when you both know ...

Hold up your hands with the number of 1/6's shown by your fingers.

Great, there are three 1/6's in 1/2.

Now, I want you each to work with a partner to discuss how you might come up with a solution to this problem:

Pretend the square is a candy bar divided into 1/6's. You and your 2 friends want to split it equally. How many pieces will each get? What part of the candybar is it ?

CHECK FOR UNDERSTANDING

Students should discuss amongst themselves how many 1/6's are showing on 1/2 of the board.

Look for a show of hands with the right answer showing with their fingers (3).

The students should be looking at the teacher and listening to her.

Students should be watching and listening to the teacher give directions.

Guided practice

After passing out a blank sheet, the teacher draws one of the boxes divided into half on the overhead or chalkboard, asking the students to think but don't say how to share the candy that is divided into 1/6's into 3 equal sizes. After students solve problems they will write how they figured it out. The teacher models the written part which says: The first thing we did was _____. Then we _____. So we think _____ pieces or _____ is the correct answer because_____.

Independent practice

The students work in partners to finish up the problem. Teacher is available if requested but may monitor the students' work to see how many of the pairs are on track.

Summary

The teacher asks the group, "What did you find out? Let's see how you figured out how many pieces." Several volunteers explain their approaches and answers. Teacher asks them, "So then, 1/3 of the candy was equivalent to how many sixths?"

Assessment

Give a similar problem to student using twelfths and fourths.

Problem Solving with Fractions

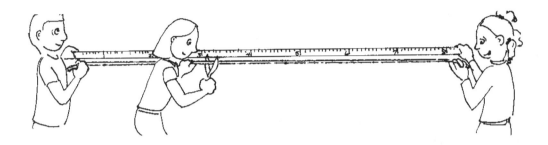

by Jennifer Emery

Fractions: a division model

Problem Solving
Students will physically divide a strip seven units long into thirds and then determine how many units long is each "third."

Materials

Math notebook ditto - Fractions in the Real World (enclosed)
Blank paper
Pencils

Set

A fraction is a part of something. For instance, we want 1/3 of six feet of rope. We divide it into 3 equal parts and get 2 feet, right? But suppose you can't divide equally, like 7÷3?

Students will take out their math notebook and tear out one sheet of paper. They will fold it lengthwise (like a fan) in one inch strips. Tear off 1 or 2 strips. Strips do not need to be straight.

Introducing

Have the students take out the strips of paper and tell them to think of the length as 7 units. Have them write 7 units on the strip of paper.

I will also have a strip of paper and call it 7 units. Next, I will flip on the overhead and draw a diagram of the strip: [7 units]

Tell them to write today's date in their notebooks and to write:
$$7 \div 3 =$$

Say the equation again and ask them to show how they would divide the strip of paper into 3 equal parts.

I will also take my piece of paper and divide it into 3 parts.

Have them hold up their strips of paper and show how they have divided it without measuring it. Then, ask for a volunteer to share how he/she got it.

Ask the students to figure out what one "part" would be equal to. Have them work it out in their notebooks and draw a diagram that would help explain their answers.

I will show the strip divided into 3 equal parts on the overhead.

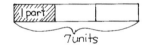

Exploring

In groups, students will draw, fold, write, and discuss to figure out how is each part. I will now circulate the room, looking for diagrams and answering individual questions. When I see most of the kids understanding the concept, I will ask for volunteers to share their answers with the rest of the class. If I feel most of the kids are struggling with the concept, I will back up, review, and work through another problem with the class. The volunteers that share their answers should be able to adequately explain their line of thinking. If they get stuck, they'll be allowed to pick a "helper" to come up and help them with their explanations and their diagrams.

Summarizing

I will ask for volunteers to come up to the overhead to show and explain their answers using the diagrams. Their answers should look something like this:

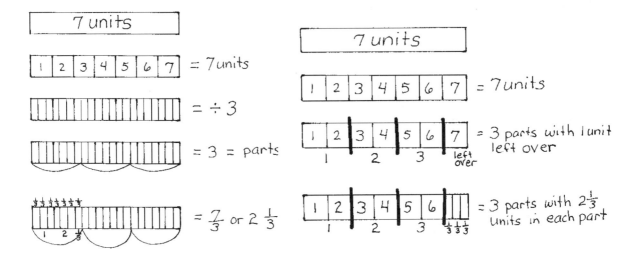

Try several other examples for paper strips: $5 \div 3 = 5/3$
$3 \div 6 = 3/6$ or $1/2$ $5 \div 6 = 5/6$
$2 \div 3 = 2/3$ $3 \div 8 = 3/8$

Have the kids make a mental summary of what they learned. Review the relationships between division and fractions, e.g. $5 \div 8 = 5/8$.
I found a bell was very important in this lesson. It was the signal for attention to the speaker - usually another student- and it kept me in control of the management while the students were free to present and discuss their work.

Homework/Assessment

I will have one person from each table come up and get enough Fractions in the Real World sheets for their group. The directions will be for everyone to complete the word problems along with diagrams in their math notebooks. Each table will also receive one transparency.

The transparency will be used by each group the next day to explain to their classmates one of the word problems on to the overhead projector. There are 7 groups and only 6 questions, so one group will be duplicating the same word problem. Let students make a final comments page to explain any changes they would make to their problems. To assess, check reasoning and diagrams in their notebooks.

Fractions in the Real World

1. Marie purchased a 5-foot length of gold chain to make identical necklaces for four of her friends. How long should she cut each piece of chain?

2. The 7 mile hike on Wilderness Trail is divided into 3 equal parts by rest stations. How long is each part?

3. Three pizzas are shared equally by 8 people. How much pizza does each person receive?

4. A roll of 300 feet of fishing line was divided equally among a group of people. Each person has 300/4 feet of line. How many people are there?

5. A small tube of window caulking will caulk 30 feet. Each side of a square window is 45/4 feet long. How many tubes much be purchased to caulk the window?

6. A recipe for baked beans calls for 3/4/ of a cup of molasses. Bob used the recipe 4 times. How many cups of molasses did he use?

Fraction Strips

by Karen Szakacs

Adding and subtracting fractions

Students will add and subtract fractions using fraction strips. They will demonstrate how to exchange strips to form equivalent fractions so they can complete the operation, e.g. 1/2 + 1/4.

Materials

Five 4" x 18" pieces of construction paper cut as shown (one each of five colors) for each student
Blank paper and a pencil for each pair of students

Transitional activities

Students have already made these fraction strips in a previous lesson (See Appendix: Fraction Strips)

Purpose

This lesson is to help you picture fractions in your mind.

Anticipatory set

Show pieces of fraction strips "We're going to use our strips to add and subtract. How many fourths equals a half?

Model activity

You're going to work in pairs to make up and solve two fraction problems using your strips. Since you have only halves, fourths, eighths, and sixteenths, you'll have to stick to problems using only these numbers. The "writer" will write a problem (e.g., write 3/16 + 1/2 on the blank sheet) for the "doer" to solve using the strips. Once the doer has solved it (show the strips then show the exchange), the writer rewrites the problem as it was solved (e.g., write 3/16 + 8/16, since the 1/2 was exchanged for 8/16). Finally, the writer writes the solution (e.g., 11/16). Then partners switch roles. Be sure to write the names of both students on your sheet since I will collect these and read some of them to the class to solve in a few minutes.

Exploring

Students make, solve, and record the solutions to two problems. Give completed sheets to the teacher.

Summarizing

Teacher chooses several problems to read aloud and have students solve. See if any students have started to solve them <u>without</u> the strips.

Assessment

Watch students solve a problem using their strips.

It All Adds Up!

by Christie J. Sonmez

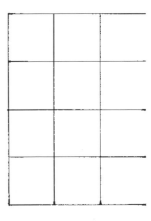

Adding halves, thirds, fourths, sixths and twelfths

This exercise will give them concrete experience in adding fractions of thirds and halves and finding the least common denominator. It will reinforce what they have been learning and will introduce them to the symbolism of adding columns of fractions.

Materials

It All Adds Up! worksheets for each student which includes a gameboard, a spinner, and a chart on which to transfer game information
Game pieces worksheet*
Construction paper *
Paper clips
Glue
* The two worksheets should be given to students the day before. Students make the game pieces by cutting them out and tracing them on colored construction paper: e.g., 1/2 is red, 1/3 is blue, 1/4 is yellow, 1/6 is purple, 1/12 is green. Color spinner to match game pieces.

Transitional

Students have had experience dividing rectangles into halves, quarters, and eighths, as well as into thirds, sixths, and twelfths. They have also been finding common denominators.

Classroom environment

Independent work with some informal talking.

Anticipatory set

I will remind the students that when I first came their class they were working on a project that I admired. They were showing thirds, sixths and twelfths on geoboard recording paper. As a follow-up, they divided a 9 x 12 rectangle into spaces that represented all of those fractions. Earlier, they had done similar work with halves and fractions of half.

Purpose

Now that the students are adding fractions, I thought it would be fun to mix the earlier exercises they had done. It would be interesting to fill in a similar rectangle with combinations of halves and quarters and thirds, sixths and twelfths. How many different combinations of those fractions can be created among all the students. Each student will be working independently, contributing one combination to the project. In order to get away from the obvious – two halves, four quarters, three thirds, etc. – the students would use the element of chance in combining the fractions.

INPUT	MODEL	CHECK FOR UNDERSTANDING
I will be giving you a gameboard, a spinner and fraction pieces. Here is the board that we will be covering with the fraction pieces. How many squares are there on the board?	Show gameboard.	Voluntary response. (12)
I will ask a student if he can find a construction paper cutout that will exactly fit one of those squares.		Student will locate a green piece.
Yes, the green piece fits in one the twleve squares. What should I call this piece, if it is one of twleve pieces?	Hold up green square.	Voluntary responses.of (1/12)
Correct. It is 1/12 of the rectangle.		
The rectangle is made up of 12 squares. If I were to cover half the rectangle, how many squares would I need to cover?		Voluntary response. (6)
Can someone find a cutout that covers six squares of the rectangle?	Place half piece on board.	Student finds half piece.

Number

INPUT	MODEL	CHECK FOR UNDERSTANDING
Yes, the red cutout covers half the board. Six squares are covered and each square is 1/12. Is there another way to express one-half.		Voluntary response (6/12)
As you discovered with your geoboard paper, there can sometimes be more than one way to show 1/2. What makes these different? The same?	Hold up two configurations.	Voluntary response (different shapes; both cover 6 squares)
We will similarly discuss the remainder of the fractions. Variation: Can someone find a piece that, if we were to put three of them down, would cover the board?		Volunteers show pieces e.g.
Having run through all the pieces, I will ask what piece they all had in common.	Hold up the twelfth.	Voluntary response
We would take a look at the spinner, noting that it is divided into 8 parts. 3 parts represent twelfths (green) and two are sixths (purple). Why did I have more twelfths and sixths?	Point to each part of the spinner as it is described.	Speculation e.g. (The smaller the fractions the more pieces we get onto our boards, the greater number of combinations.)
I will demonstrate how the spinner works. What color did it land on? What fractional part is that?	Demonstrate spinner.	Volunteer names color and part.
Let's try this once all together. I'll spin.		

Guided practice

I will spin the spinner and call out the color and fraction name. Each student will find the corresponding piece and lay it on his board. We will continue play until our boards are filled up. Should we spin a piece that is too large to fit in the remaining space, we must spin again. When our boards are filled in, I will demonstrate and say that "this fraction" and "that fraction" and "the other fraction" all add up to one whole.

Independent practice

The students will now have the opportunity to fill their own boards, following the dictates of their own spinners. When their boards are filled in, they will glue the pieces down. If they finish early, they should sit quietly, observing the other results or spinning again to see if they get another combination.

Closure

We will compare our results. Hopefully, each will be different. As we discuss our results, the students will chart the number of halves, quarters, thirds, sixths and twelfths they put on their boards. Fraction by fraction, we will talk about how many we used. Example: Who spun blue? How many? Put the number over the four on the chart. How many twelfths did you cover with fourths? Put that number over the twelfths across from the fourths. You didn't get any? Put down a zero. We will total the number of twelfths we charted
e.g. $0/2 + 0/3 + 2/4 + 0/6 + 6/12 =$
$0/12 + 0/12 + 6/12 + 0/12 + 6/12 = 12/12 = 1$

Assessment

I will know that the students have gotten it when their boards are filled up with construction paper fractions and they have charted their fractions, as the pieces they had spun and as twelfths. If they come up with a total of twelve twelfths and equate that total with one, they've gotten it.

Follow-up

We speculated how many squares we would need to employ an eighth. A rectangle of twenty-four squares is another possibility. Immediately, the task is to compare all the combinations we had created as a class and discover how many unique ones we had. A challenge question: How many more could we create if we weren't restricted to using the spinner.

Sources
Arithmetic Teacher: "Fraction Squares" – Jan. '85, p. 43; "Fractions with Fraction Strips" – Dec. '84, p. 4-9.

It All Adds Up!

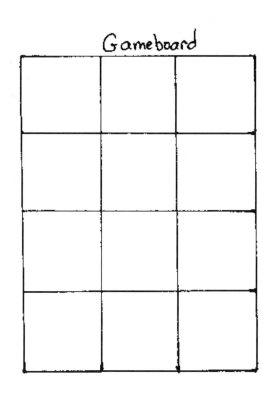

Gameboard

How many?

$$\frac{}{2} = \frac{}{12}$$

$$\frac{}{3} = \frac{}{12}$$

$$\frac{}{4} = \frac{}{12}$$

$$\frac{}{6} = \frac{}{12}$$

$$\frac{}{12} = \frac{}{12}$$

$$1 = \frac{}{12}$$

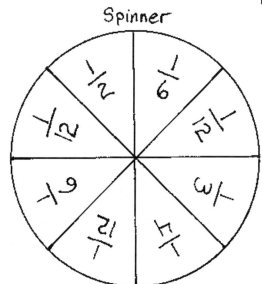

Spinner

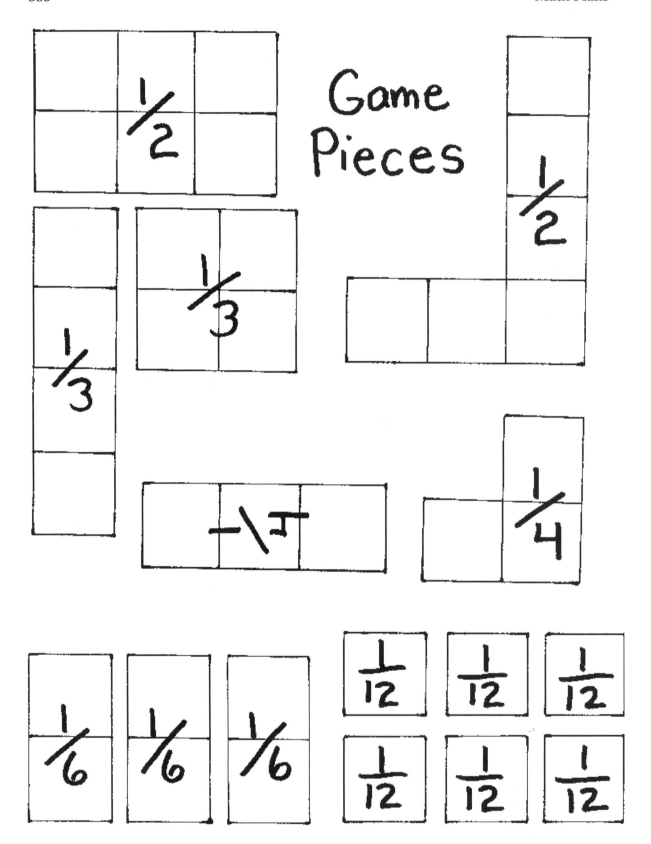

"Tenths" Using Base 10 Blocks

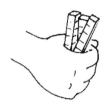

by Susan Frost

Decimals:tenths
Students will represent tenths using "longs" from their base -10 materials.

Materials

A base-ten "flat," ten "longs" and two copies of the Decimal Paper (Appendix) for each child.

Transitional activities

This is the first lesson of a unit introducing decimals and decimal place value. The children have been reviewing subtracting large numbers and subtracting across zeros. Besides recognizing the decimal point in money, they have not been introduced to decimals.

Classroom environment

I will allow time for the class to play with the base ten materials for about ten minutes before I begin instruction. At this time, noise will be allowed as the students explore the manipulatives. However, during instruction, the class should be attentive and talking noise will not be acceptable.

Anticipatory set

I will tell the class that today instead of their textbooks we are using manipulatives and a game to learn decimals.

Purpose

I will tell the children that the base-ten materials and worksheet will show them what "tenths" look like.

Input

I will start the instruction by describing how each base-ten stick is part of the base-ten flat. Ten longs make up one flat. Each flat is a whole, each long is a "tenth." When we move on to the worksheets, I will ask the kids what part of the square is each column.

Model

I will demonstrate with the base-ten materials how one long is a tenth of the flat. I will also draw a square on the board consisting of ten columns. As I say a fraction, such five-tenths, I will color five columns. I will also write the symbol and the written way to say five-tenths by the square (0.5). I will demonstrate with about five decimals.

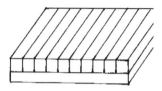

Check for understanding

I name the decimals, the students will make the decimals with their manipulatives. I will look around the room to see if they are showing the correct decimal.

Guided practice

I will write nine decimals symbolically (e.g. 0.2) on the board and the students will copy each decimal above a square. They will then fill in the corresponding amount of tenths. When they complete this exercise, I will dictate nine decimals for them to complete the second worksheet. Again, they will write a decimal I say above each square, and fill in the corresponding tenths columns.

Closure

I will fill in the decimals that I had dictated on overhead grids as students check their work.

Independent practice

Because this is the first time the class will work with decimals, I will not assign independent work. We will follow through in a later lesson.

Follow-up activities

In the next lesson, I will use the "units" from the base -10 materials to show hundredths. Have students show hundredths on grids as before and compare hundredths with tenths e.g. .6=.60 How can you prove this? .6 > .45 Why?

Assessment

Have student explain whether .45 is less than, greater than than, or equal to .6 using base-10 materials.

Multiplying Decimals

by Jennifer Emery

Students will solve problems involving multiplying decimals using either pictures, diagrams or "number crunching." They will explain their processes.

Materials

Two transparencies for each pair or small group
Felt pens for transparencies for each pair or small group
Overhead projector
Scratch paper

Transitional activities

Students have already been introduced to these three models of solving the problems. This is one of a series of lessons to allow the students themselves to explain and clarify their understandings to each other using the models.

Classroom environment

Lots of talking in groups during exploring. Involved and attentive listening when someone is presenting at the overhead projector during summarizing. Signal for attention is a bell. The bell will probably be used several times during summarizing to regulate the discussion.

Anticipatory set

As students enter the room they are directed to the overhead screen in front which shows a greeting and directions to begin the first problems in their groups. One group representative gets two transparencies and a pen to record the group's work. "Solve this using either pictures, diagrams or number crunching: .5 x 1.2."

Exploring

Students work in pairs or groups on transparency. Bell signals the end of exploring.

Number

Summarizing

"OK., I need a volunteer to show us what this looks like." Choose volunteer. "Are you going to use pictures, diagrams or number crunching?" Student answers, "Diagram." (Note: This is the text of an actual lesson.)

INPUT	MODEL	CHECK FOR UNDERSTANDING
Volunteer 1 We made 1.2 first (points) Then we made .5 here.	Puts transparency on overhead. 	Teacher: Does that side show .5 or 1.5? (Student is puzzled.) Teacher asks class, "Could someone help her?"
Volunteer 2 This side is .5 The longs are 1 unit long and the cubes are 1 tenth.	Voluteer 2 puts transparency on overhead. 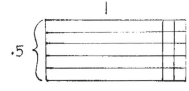	Teacher: Anyone get a different answer? A student raises hand: I don't see where you got the .5 from.
Student explains that there are 5 longs and each long is .10.		But you wrote that a cube = .10.
(Looks at acetate.) That means that altogether when you added up all the cubes it equals .10.		Is there another way to explain it? Someone have another way?
Volunteer 3 I showed the whole thing first. The I took half of it because that's what .5 is. So the answer is .6.	Puts acetate on overhead 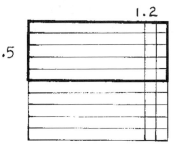	Teacher: How many understand this explanation? (Lots of hands go up.) Who has another way? What will you use?

INPUT	MODEL	CHECK FOR UNDERSTANDING
<u>Volunteer 4</u> Number crunching.	1.2 <u>x .5</u> .60	(Several hands go up.) Teacher: There are some questions. Do you want to call on someone?
Anne.		Anne: How did you know where to put the decimal point?
You count the number of places behind the decimal point. That's how many are in the answer.		Teacher: Other questions? (none) Here's our second problem. Remember that everyone in your group should be able to understand the explanation your group representative gives.

Exploring

Groups work on the problem .6 x 3.2 for 10 - 15 minutes as teacher observes for participation and understanding. Students in several groups express disbelief in their group's answer. It doesn't make sense to them that the answer in this multiplication problem could be smaller than 3. Several borrow calculators from the math closet to check it.

Summarizing

Signal for attention: Bell. (Remember to use your signal to call the class to attention whenever the discussion becomes unproductively loud.) "We need some good explanations for this one. Who will start us off?"(volunteer) "What will you use?"

Number

INPUT	MODEL	CHECK FOR UNDERSTANDING
<u>Volunteer 1</u> A diagram. First, there are 3 stacks of .6. These are the first 3 bigger ones, the longs. Each long is .1. That's why there are 6 of them in each. Then comes the littler ones. There are 12 of them because it's 2 x 6. Each cube is .01. So 12 cubes would be .12. The answer is 1.8 + .12 = 1.92	Puts acetate on overhead. 	
		Loud discussions. (Bell) (Many hands go up.) Teacher: Would you like to choose someone for a question?
Greg.		Greg: This is our group who did this one and we all agree it's right.
	Volunteer 1's group stands up front with him	Teacher: Why don't we have all four up in front and see if we can answer questions from other groups.
Carla.		Carla: We got 3 for our answer because those 12 cubes when you put them end to end make a long and 2 cubes left over. So that's 1.2. Then we added 1.8 to 1.2 and we got 3.
(Silence from volunteer 1's group as they are still processing what was said.)		Teacher (after some wait time): Does everyone agree with Carla that 12 cubes makes 1 long and 2 cubes left over? (Many hands go up.) So do you think the answer is 3?
		Ellie: I agree with Carla but when I multiplied this on the calculator I got 1.92.

INPUT	MODEL	CHECK FOR UNDERSTANDING
Volunteer 1: I think I can explain it. If you do what Carla did, you get another long and 2 cubes. Now, there are 19 longs and 2 cubes. Nineteen longs doesn't make 3 flats. It doesn't make even 2 flats. The answer had to be less than 2.		Teacher: So, then, what does that mean for your answer?

Closure

Everyone take your seats and write on the next blank page in your math notebook what you thought the answer was. Did you change your mind - yes or no - and why.

Assessment

.3 x 3.5 Have students solve this using a diagram, picture or number crunching and write a paragraph about why they think it's right.

Fractions as Rates

by Darcy Cooper

Fractions as rates

The main objective of this lesson is to have the students learn about rates as a context for exploring fraction equivalents.

Materials

Newspapers - to find rates
Beans - as counters
Coins - as counters

Transitional activities

This is the first in a series of lessons on rate. The students have had other models of fractions before e.g. strips, pies, geoboards.

Classroom environment

The children will be asking questions and volunteering answers at the beginning - one at a time. However during some of the activities they will be working in groups of four at their tables and I will expect them to speak quietly with each other.

Anticipatory set

I will begin with a discussion of rates, such as how baby-sitters are paid by the hour and the speed limit. I will write examples of rates on the overhead. Next, I will ask them for examples.

Purpose

To have the students find their own examples of rate, in order for them to understand better what a rate is.

Introduction

I will have the students look through newspapers and find rates to use as examples. They will work in groups at their tables looking through the newspapers to find rates. They will tell them to me and I will write them on the overhead and we will discuss them e.g. 2 men's tee shirt/6 dollars; 4 tires/$239; 1 phone/$259.99.

Guided practice

The next step will be to set up a base rate and then to generate a series. I will present them with this problem.

"I found a real bargain at a flea market. I bought a tremendous box of candies. I want to sell them in packages of 5 for 2¢. If you want to buy 10, how much money do you have to give me? If you want to buy 15?"

Show beans and coins on overhead:
We will discuss this for a while. Next, I will have them represent the problem with the beans and coins that they have at their desks. The beans will be the candy and the coins will be the money. I will have them build a series of these models so they will be able to see what the series looks like.

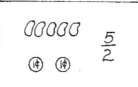 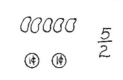

As we discuss I will write the series on the overhead as fractions: 5/2; 10/4; 15/6; 20/8; 25/10...

I will ask the kids to find how much six packets would cost and have them explain why.

Independent practice

Next I will give them another problem to work in groups. "I found a vending machine that dispenses peanuts for 4¢. I got 7 peanuts every time. After 6 times, how many peanuts have I got and how many pennies did I spend? The base rate is 7 peanuts / 4¢. Work in groups and build a model and record the rate on paper. 7/4¢, 14/8¢, 21/12¢, 28,16¢, etc.

I will have them discuss their results with the whole class.

Closure

Ask the students if they notice patterns in the numbers and try to bring into focus the fact that the numbers are equivalent fractions.

Follow up

I will assign one or two problems as homework e.g. Sandy runs seven miles a day in the mornings. How far does she run in six days? Show this and find how far she would run in eleven days.

A sale in a clothing catalogue offered two dresses for $38./ The normal price is $224 per dress. At the sale price how much do four dresses cost? How many sale dresses could you buy for $224?

Source

Seeing Fractions by Rebecca B. Corwin, Susan Jo Russell and Cornelia C. Tierney. California State Department of Education, 721 Capitol Mall, Sacramento, CA P.O. Box 944272, Sacramento, CA 94244-2720. 1990

Ratios - Cubes to Tiles

by Holly Whalen

Making ratios and recording them three different ways e.g. 2 to 3; 2:3; 2/3. In this lesson students will make ratios by grabbing a handful of cubes and a handful of tiles and recording this ratio the third way - as a fraction.

Materials

Cubes and tiles, pencils and paper for every group.

Transitional activities

Students have done a similar activity and recorded the second way e.g. 2:3.

Classroom environment

The students will talk among themselves in groups as they make the ratios and record them.

Anticipatory set

(Group monitors are directed to come up to get a bunch of cubes and tiles for their groups.) We are going to show some ratios again using the cubes and tiles. Remember how we recorded them? (Write on board: 2 to 3; 2:3; 2/3).

Number

Introducing

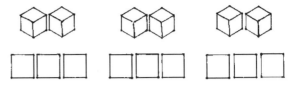

Call a volunteer up to grab a handful of tiles and another of cubes and place them on the overhead. How many do we have? (6 cubes, 9 tiles) How did we write this last time? (6:9) Today, let's write it this way: 6/9. Could we write this a simpler way? (Volunteers come up to reorganize the materials two cubes with three tiles) What does this show? (For every 2 cubes there are 3 tiles). So how would you record this? (2/3) Suppose we did this lots of times until we had 100 cubes and we wanted to know how many tiles. Write this on the board:

$$\frac{2}{3} = \frac{100 \text{ cubes}}{? \text{ tiles}}$$

Would that number be bigger or smaller than 100? (Bigger) What would the number be? (Ask several volunteers to explain what they think.) In your groups you will be answering this question and another one. Suppose you got 100 tiles. How many cubes would that be? Write this on the board:

$$\frac{2}{3} = \frac{? \text{ cubes}}{100 \text{ tiles}}$$

Exploring

Students use manipulatives, draw pictures, and discuss as they record their answers. 2/3 = 100 cubes/150 tiles; 2/3 = 66 cubes (approx.)/100 tiles

There will be a lot of discussion about the second one because the numbers do not divide out evenly and so the answer (66 or 67) is not exact.

Summarizing

In the first one, many of you said the number of times would be more than the number of cubes. Why did you say that? What did you get for the number of tiles? How did you get that number? (Get as many different explanations of the first one as there are in the class.)

For the second problem, ask if they think the number of cubes should be less or more than the number of tiles. (Less) Why? Continue as in the first problem.

Assessment

The ratio of boys to girls in our class is 18/12. Draw this as a picture. If you think this ratio can be written more simply, arrange the boys and girls in groups to show this and write the simpler ratio.

Introducing Percents

by Michael Knofler

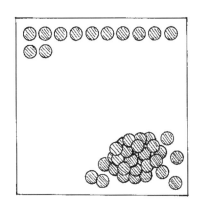

Concept of Percent

Students will use 100 pennies and base - 10 materials to concretely show percent.

Materials

Rulers, one hundred pennies (one set) and base ten "flats" and "longs" for everyone.

Transitional activities

Percents have a direct relationship to fractions and decimals. These sixth grade students have already studied these areas, both conceptually and symbolically. Ms. Gibson has assigned them word problems through POWs (problem of the week) and class work problems dealing with fractions and decimals.

Classroom environment

I expect to teach this lesson to small groups (four to six students) in an open and cooperative setting. I expect minimal noise, mostly relating to questions, answers and comments regarding the lesson. I will tell the students to use their 12" voices when working with one another.

Anticipatory set

I will open the lesson with a discussion on how percents are involved in our daily lives. I will present clippings from newspapers and magazines that show percents. I will talk about sales reports, tests, batting averages, etc...I will encourage an open discussion about what type of percents the students deal with in their lives and what they know about percents.

Purpose

I will tell the students that percents are a part of our daily lives and need to be understood to comprehend certain aspects of our society. Then, I will again pick up the newspaper and magazine clipping to emphasize that point.

INPUT	MODEL	CHECK FOR UNDERSTANDING
After I have explained the purpose of this lesson, I will hand out the base ten flats.	Pass out base ten flats. Point to the ten vertical and ten horizontal squares and ask how many squares are in the base ten flat.	I will ask the children to raise a flat hand if they understand how the flats equal 100 squares and a fist if they do not. If they do not understand, I will have them multiply 10 x 10 and explain that there are ten squares in ten rows and that by multiplying 10 x 10 they are figuring out how many total squares there are. If they understand, I will go on, if not I will work with them until they do.
I will tell them that each square equals 1%. I will explain that percent means 'one part in one hundred'.	If someone does not understand, I will explain why one square equal 1%.	Ask for hands (as described above) to CFU why each square equals 1%.
I will then ask everyone to cover the first horizontal row of 10 with their ruler.	I will also cover one horizontal row with my ruler. I will explain to anyone who understands why they are covering 10%.	I will look to make sure everyone is doing what I asked, and then I will call out for the answer to what percent we are covering. I will ask for hands to see if everyone understands.
I will continue to ask the students to cover various rows.	I will cover the various rows with the students and explain to anyone who does not understand why a certain percent is the amount of percent they are covering.	I will continue to use call outs for the students to tell me what percent we are covering.

INPUT	MODEL	CHECK FOR UNDERSTANDING
I will then tell them to cover various percents and explain why they need to cover a certain amount of the flat.		I will look over their shoulders to see if they are covering the correct percents.
Next, I will tell the children we will move onto money.	I will present 100 pennies up as ten rows of ten, exactly the same as the base ten flat.	I will call out for students as to where I should put the four rulers to make the five splits.
I will ask them to split the pennies in five equal ways.	I will place the rulers where the student points.	I will call out for answers and have the students explain their answers. I will ask for a show of hands to CFU.
I will ask them what percent is in each of the five groups.		I will continue to call out and ask for hands to make sure everyone understands. I will also ask each child individually to answer at least one question out loud.
I will continue to ask them to split various percents with the pennies. I will also ask them to convert the percents into monetary forms (i.e. - 50% equals how many dimes? quarters? nickels?).		
	Next, I will show them the ten piece FLU's.	
Explain to anyone who does not understand why covering the first two pieces equals 20%.	I will cover 20% and various other percents with my ruler.	

Independent practice

I will tell the students to get their base ten flats and pair off with one another. I will then group them as "A" partner and "B" partner. I will then tell them that the "A's" will ask the "B's" to cover five percentages that the "A's" will choose. Then they will reverse. I will ask everyone if they understand by giving me a show of hands. I will look over the groups to make absolutely sure they understand both the exercise and the percent fundamentals.

After they have completed this lesson, I will have them do the same task with their ten piece FLU's. I will monitor this lesson the same way.

Closure

I will ask the groups the following questions:
1. What is percent?
2. What did you learn about percent that you did not know before this lesson?
3. How do you think knowledge of percent will help you?

I will direct a group discussion and get everyone's input. If someone is not participating, I will ask them to tell me what they learned today to try to ease them into the discussion.

Assessment

Ask students to answer the following questions in their journals:
1. ($.25/.$50/$.75/$90) is what percent of one dollar?
2. If I cover (3/5/7/9) sections of the ten piece (long), what percent am I covering?

Fun with Percents

by Lori Souza

Visual concept of percent

Students will guess what percent of a card is colored and what percent is not.

Materials

Money (i.e. 100 pennies, 10 dimes, 4 quarters) 20 percentage cards, article clippings from newspaper and food labels that deal with percents.

Transitional activities

Percents are important since they relate to fractions and decimals. Percents are introduced after students understand fractions and decimals and have had experiences with ratios.

Classroom environment

Groups of 4 -6
Low Noise level (6-inch voices during group work)

Anticipatory set

Ask students what is a percent and where have they seen percents. Show articles from the newspaper that use percentages, such as test scores or advertisements. Additionally, show students some food labels that have percents written on them. Follow up with an informal definition (part of a hundred, portion or fraction).

Purpose

Percents affect them in their daily lives.

Input

Explain money analogy with percents (i.e. one quarter is 25% of one dollar).

Model

Illustrate a wide range of percents, such as 50%, 90%, 5%, 100%, 99%, and 1%. Show students different portions of money using 100 pennies, 10 dimes and 4 quarters.

Check for understanding

Choral responses — Ask questions such as : If I take away 3 quarters what percent did I take away and what percent remains? Ask similar questions using pennies, dimes, and quarters.

Input

Further illustrate the percents by presenting a meter stick as a model.

Model

Cover part of the meter stick and ask children to estimate the percent of the meter stick that is covered.

Checking for understanding

Call outs — ask students to estimate what percent of the meter stick is covered and what percent of the meter stick is not covered?

Guided practice

Present the card game. This activity is related to area. The card game is a stack of twenty cards with the B side down. On the A side, specific percents of the card are colored blue (0%, 10%...100%) and the remainder (100%...0%) red.

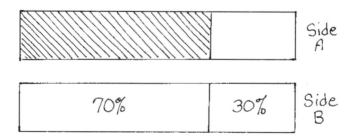

Number

Show the students how to play the game: demonstrate a couple of times or until they understand what to do. Randomly pick a card and try to estimate the percent. If the guess is wrong, then the card is not kept.

My estimation would be 50% red and 50% blue. I turn the card over and discover I am correct. I keep the card.

My estimation would be 20% red and 80% blue. I turn the card over and discover I am not correct. (It is actually 25% red and 75% blue.) I do not keep the card.

Independent practice

In pairs, have the students take turns trying to win cards by looking only at side A and trying to predict the percents shown on side B. The student picks up the card on the top of the pile and tries to predict what percentage of the card is colored blue and what percentage of the card is red. If they estimated correctly, they get to keep the card. The one with the most cards at the end of the game is the winner.

Closure

Ask these questions.
1. What is a percent?
2. Why do you need to know percents?
3. How does money relate to percent?
4. If I take 20% of something what percentage remains?
5. If I take 45% of something what percentage remains?
6. If I take 0% what percentage remains. If I take 100% what percentage remains?
7. What pattern about percent did we discover? (The sum of the covered and uncovered portions will always total 100%)

Assessment

See how good they are at guessing the percent.

50%	75%	90%	80%	70%
			20%	30%
		10%		
50%	25%	15%	35%	45%
	5%			
40%	95%	85%	65%	55%
60%				

Spinners

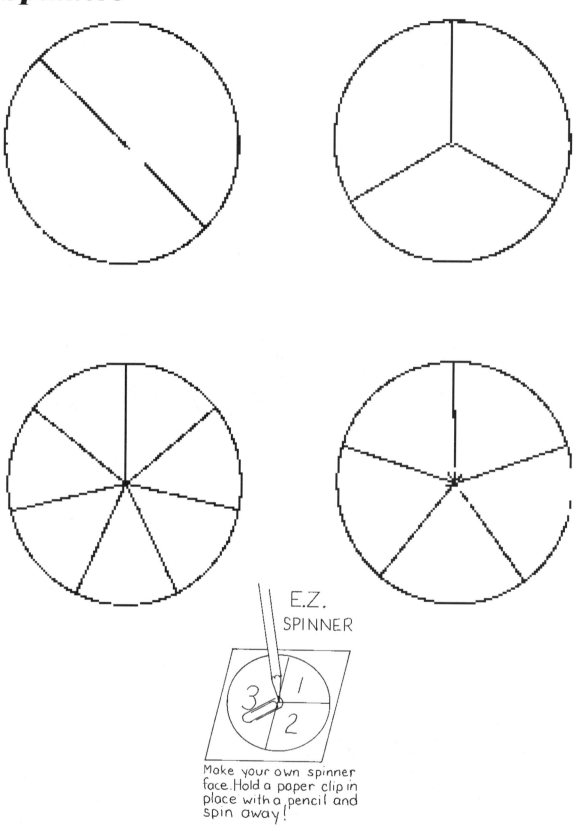

Geoboard Template

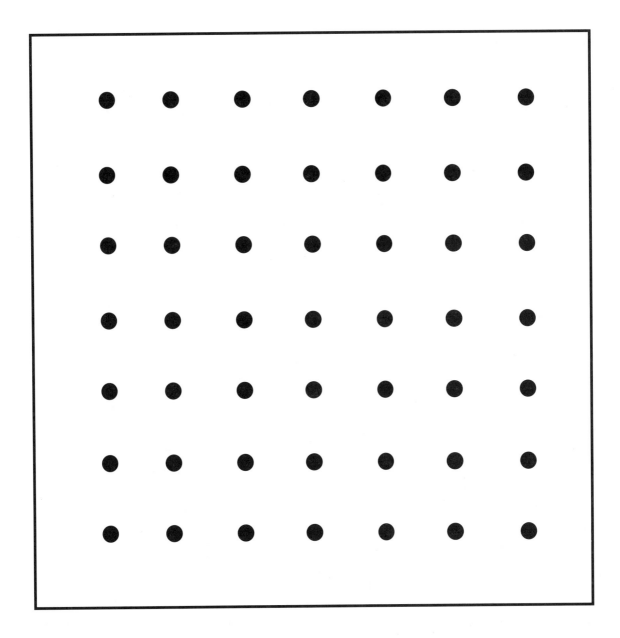

Geoboard Recording Paper

Centimeter Dot Paper

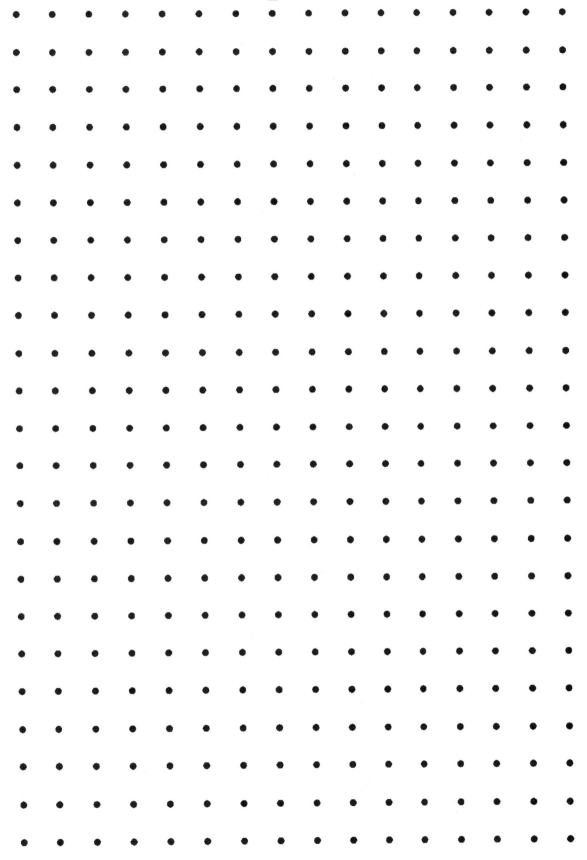

Isometric Dot Paper

Inch Grid Paper

Half-inch Grid Paper

Appendices

Quarter-inch Grid Paper

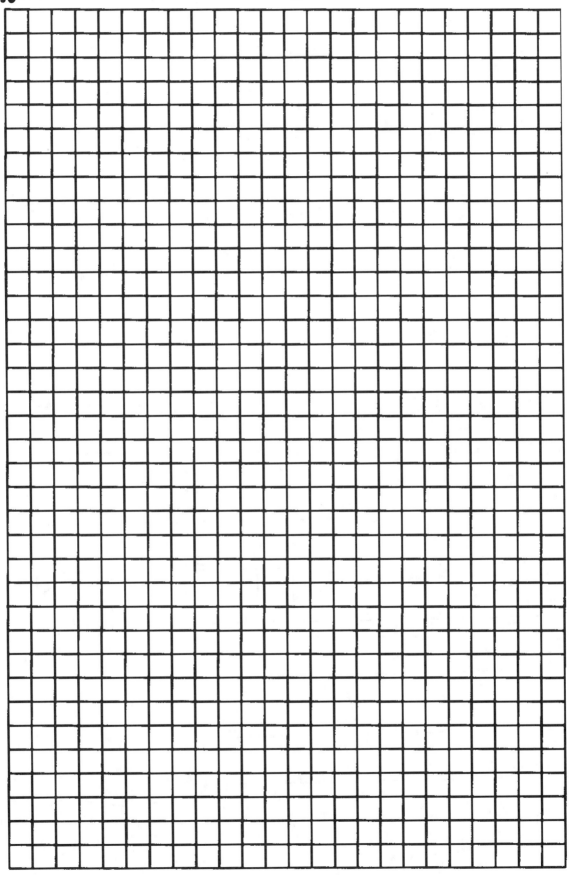

Rulers

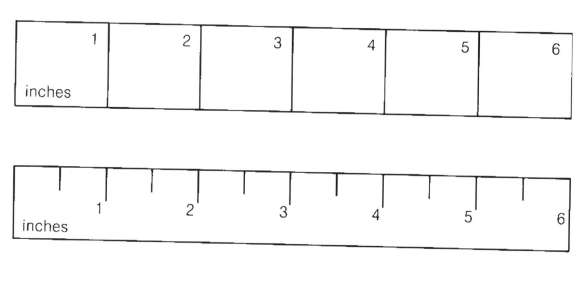

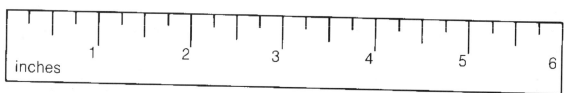

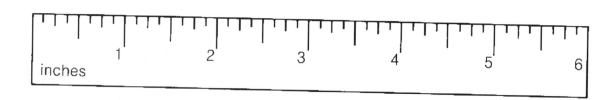

Appendices

Pattern for Base-10 Materials (inches)

Base-10 Mat

Appendices

Hundreds Square

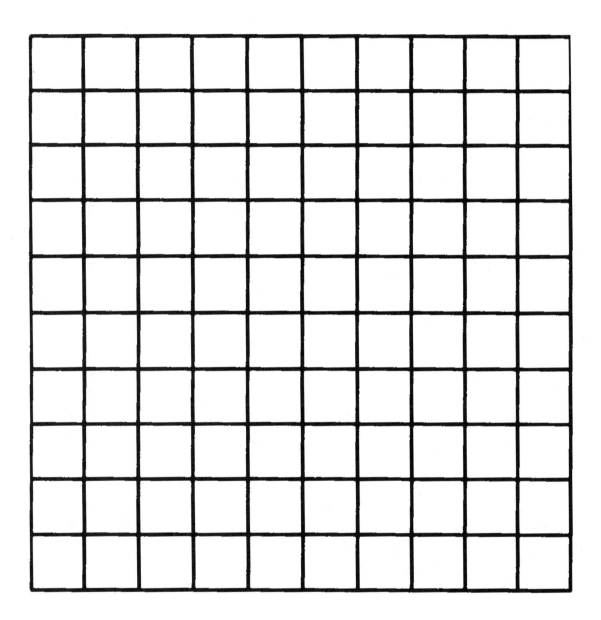

Hundreds Charts

0	1	2	3	4	5	6	7	8	9
10	11	12	13	14	15	16	17	18	19
20	21	22	23	24	25	26	27	28	29
30	31	32	33	34	35	36	37	38	39
40	41	42	43	44	45	46	47	48	49
50	51	52	53	54	55	56	57	58	59
60	61	62	63	64	65	66	67	68	69
70	71	72	73	74	75	76	77	78	79
80	81	82	83	84	85	86	87	88	89
90	91	92	93	94	95	96	97	98	99

Appendices

Fraction Strips

whole	half	fourth	eighth	sixteenth
				sixteenth
			eighth	sixteenth
				sixteenth
		fourth	eighth	sixteenth
				sixteenth
			eighth	sixteenth
				sixteenth
	half	fourth	eighth	sixteenth
				sixteenth
			eighth	sixteenth
				sixteenth
		fourth	eighth	sixteenth
				sixteenth
			eighth	sixteenth
				sixteenth

Fraction Models

A Teaching Self-Evaluation
by Caroline Miller

A self-evaluation of a lesson in fifth-grade using manipulatives (base-10 materials consisting of beans, bean sticks, and bean rafts), a dice and numeral cards. This student teacher is using manipulatives for the first time in a class that does not use groupwork.

I wanted the children to show me their knowledge of addition and subtraction and the addition part was a SIN: They knew immediately that the winner would just have to know in what column to put the most amount of beans and they were knowledgable with this skill. Most of the time, the game was tied and, unconsciously, I thought about a tie-breaker based on speed - the one who gets the answer first wins. That thought quickly died. Emphasizing cooperation versus competition made **me** feel better but I don't think these kids felt the same way, especially with my comment, "Well, look how good your team is doing together. You were all as correct as you could be!" The subtraction part produced different results, however. The children didn't always come up with the same answers. For example, when I asked for the largest number, some of the kids would create the smallest number by starting out with the lowest numbers in the higher place values, especially in the hundreds place. I felt some of them needed to listen better.

Yes, the students did accomplish what I wanted. I thought the addition and subtraction would be easier for them than it was. They all thought it would be easy, too, and they were all full of confidence. Adding up the beans was no problem and making the biggest addition problem was not a struggle, but a few times there were a couple of children who produced an answer and forgot to carry over the beans (10) into the hundreds place. When asked to make the smallest addition problem, some forgot and made the biggest. I was glad to see the checker catching these mistakes, at least some of the time. Some kids caught on right away to the idea of the subtraction. But some made the smallest number when they were supposed to make the largest.

The social skills were an important part for me because I wanted to see how cooperative learning would work with a group of fifth-graders who had little, if any, experience in this area. I went over the rules at the beginning of the first lesson and had them read these rules and explain their individual roles in their own words. Then we talked about why a group of people need to follow some behavior guidelines. Somehow, during the actual game, they forgot and began making rude, fifth-grade remarks.

The strengths of this lesson were, first, children who had little experience with manipulatives in the upper grades would get to "feel" their answer. In actuality, though, the answer "in beans" was sometimes left as it was originally constructed without making exchanges and taking off the unnecessary beans. When more challenging questions were asked with subtraction and finding the smallest number for the answer, they really had to think and they used trial and error. Most began with the mental computations but one boy immediately went to the beans. A girl, Sarah, grabbed the paper and pencil and wrote the problem out (!), or at least what she thought the best selections of numbers would be. This seemed typical for someone who probably hasn't worked with physical objects. I directed her back to the beans and she did fine. I think that transfering from the symbolic work she is used to, to the concrete work of this activity is backwards but she didn't know the concepts as was shown in her symbolic work.

These kids were on task most of the time but then again, it was a card/bean game instead of their usual teacher-directed lesson which they are used to four/fifths of the time. I did notice some grabbing, some "me first" attitudes, and a sense of competitiveness. For instance, in one group the checker would say to the only girl in the group, "Ha Ha! You got it wrong! Wa!Wa!Wa! Wa! WA!" that sort of thing. I would quietly ask everyone to remember their conduct rules and then they were OK. When I got an "I don't understand" look from some group members I said that I was sure that someone in the group did remember the directions and gave it to the groups to work out, or I would rephrase the problem and give it back to them. I found that I really had to think to remember all of my own rules.

Some things that I would do differently are that I would not have put those three boys with that one girl. The boys ganged up on the girl, at first. Their little fights and rude remarks were distracting. Obviously, they need to learn how to get along. Sarah came back later with comments like,"Give me five, SUCKAAHHH!" when anyone got points (especially the boys). They became like little "Rockys" after winning a fight. Another change I would make is to have the rules for groupwork in large print for everyone to see. That would have been helpful. As for what I would keep the same, I would use this game again, even though it was only addition and subtraction. Some of the fifth-graders were not able to make the smallest or largest subtraction problem given six numbers, or they didn't compute the right answer even though they had beans. I tried to make them think for themselves which is helpful but at the end when I wouldn't give them the answer, I heard "You use the biggest numbers in the hundreds and ten's columns, stupid," -type comments among the students. How do you get kids to stop the insults??

Was I satisfied with the lesson? Yes, but the rude comments mentioned earlier were unnecessary. Maybe I was not firm enough? They do not talk to to each other around their regular teacher. I could see what a confidence booster having jobs/roles can bring children. They really felt important and, at the very beginning, they were performing and fulfilling these roles, almost with dignity and respect, something I have not witnessed in these fifth-graders before.

Math Plans

I think these kids need exercise in learning how to listen and communicate with each other which are highly suggested in many of the books I have read this semester. I do have some neat ideas which I have copied from a number of books. In my own class, I would definately begin the year with "learning the skills of group work."

One boy, Ryan, had trouble with zero(0) which actually is a number but is represented as no beans. He wanted to put something there. He placed a bean there once and I didn't say anything. The "checker" had the most difficult job because he had to check everyone's procedure and answer. After a while, I had them switch roles so that everyone could play the game.

The discussion at the end showed me that they would love to come back for more.

Group Processing

What skills were you practicing?

_____

_____

Names of participants   Roles

1 _____

2 _____

3 _____

4 _____

What did your group accomplish?

_____

_____

_____

What helped you get it done?

_____

_____

What got in your way? _____

_____

from Johnson and Johnson, 1984

Trouble-shooting an Exploring Lesson

Introducing
1. Was your lesson objective clear in your mind?
2. Were the materials ready?
3. Did you "settle them down" to listen?
4. Did you "hook" them with your anticipatory set?
5. Did you **explain** the activity clearly, **model it** and **check for understanding (CFU)?**
6. Did you explain or review the **rules** of behavior for this activity and check for understanding of the rules with the children?
7. Did you form the **groups**, explain or review the **roles** of each group member and check for understanding of the roles with the children?
8. Did you establish a **signal** for getting their attention at the end of exploring?

Exploring
1. Did you take on the role of **facilitator** or were you still trying to "teach?" (How many times did you interrupt exploring to say something to the whole group? Did you turn exploring into a kind of guided practice?)
2. Did you **observe and informally assess**?
   Who is understanding? Who is leading the discussion? Are the quiet ones participating? Are the non- or minimal-English-speaking students participating? What students get along? Which ones do not? Who are the "social outcasts" and how could you involve them next time? Are the jobs working? Are the acts of common courtesy being observed? Are the boys doing all the work? Are the girls doing all the work?
3. Did they respond to your signal for attention?

Summarizing
1. Did your **signal** work? Were the children attentive to you or to the children you chose to speak?
2. Did you make it **safe** to share thoughts and results that might not be correct?

GENERAL SCORING RUBRIC FOR OPEN-ENDED QUESTIONS
Used for 12th grade questions in the California Assessment Program

Demonstrated Competence

Exemplary Response...Rating = 6
Gives a complete response with a clear, coherent, unambiguous, and elegant explanation; includes a clear and simplified diagram; communicates effectively to the identified audience; shows understanding of the open-ended problem's mathematical ideas and processes; identifies all the important elements of the problem; may include examples and counterexamples; presents strong supporting arguments.

Competent Response...Rating = 5
Gives a fairly complete response with reasonably clear explanations; may include an appropriate diagram; communicates effectively to the identified audience; shows understanding of the problem's mathematical ideas and processes; identifies the most important elements of the problems; presents solid supporting arguments.

Satisfactory Response

Minor Flaws But Satisfactory...Rating = 4
Completes the problem satisfactorily, but the explanation may be muddled; argumentation may be incomplete; diagram may be inappropriate or unclear; understands the underlying mathematical ideas; uses mathematical ideas effectively.

Serious Flaws But Nearly Satisfactory...Rating = 3
Begins the problem appropriately but may fail to complete or may omit significant parts of the problem; may fail to show full understanding of mathematical ideas and processes; may make major computational errors; may misuse or fail to use mathematical terms; response may reflect an inappropriate strategy for solving the problem.

Inadequate Response

Begins, But Fails to Complete Problem...Rating = 2
Explanation is not understandable; diagram may be unclear; shows no understanding of the problem situation; may make major computational errors.

Unable to Begin Effectively...Rating = 2
Explanation is not understandable; diagram may be unclear; shows no understanding of the problem situation; may make major computational errors.

Unable to Begin Effectively...Rating = 1
Words do not reflect the problem; drawings misrepresent the problem situation; copies parts of the problem but without attempting a solution; fails to indicate which information is appropriate to problem.

No Attempt...Rating = 0

Bibliography

References for Chapter 1

Burger, William F. and Shaughnessy. "Characterizing the Van Hiele Levels of Development in Geometry" *Journal for Research in Mathematics Education* , (1986)

Burns, Marilyn. *About Teaching Mathematics*. (1992) New York: Cuisenaire Company of America.

California State Department of Education. *Mathematics Model Curriculum Guide* (K-8). (1987) Sacramento: Department of Education.

Hunter, Madeline. *Improved Instruction*. (1976) El Segundo, CA.: Theory Into Practice (TIP) Publications

Inhelder, R., and Piaget, J. *The Early Growth of Logic in the Child: Classification and Seriation*. (1969) New York: Norton.

Nunes, Terezinha, Analucia Dias Schliemann, and Carraher, David William. *Street Mathematics and School Mathematics*. (1993) New York: Cambridge University Press.

Mansfield, Helen. "Projective Geometry in the Elementary School" *The Arithmetic Teacher*, March 1985.

Meyer, Carol and Sallee, Tom. *Make It Simpler*. (1983) Menlo Park: Addison-Wesley Publishing Co.

Parham, Jaynie L. "Meta-Analysis of the Use of Manipulative Materials and Student Achievement in Elementary School Mathematics" *Dissertation Abstracts International* 44A (July 1983): 96, pp. 53,54.

Pratton, Jerry and Hales, Loyde W. "The Effects of Active Participation on Student Learning" Journal of Educational Research, March/April 1986.

Stockmeyer, E.A. Karl. *Rudolph Steiner's Curriculum for Waldorf Schools*. (1982) Michael Hall, Kidbrooke Park,Forest Row, East Sussex,
RH18 5JB UK: Steiner Schools Fellowship.

Math Plans

Books for teachers

Geometry

Fouke, George R. *A First Book of Space Form Making*. Geobooks, Typesetting by Tapeset, San Francisco, 1974.
A 64-page pamphlet that gives instructions for constructing polygons and polyhedra using a compass, ruler and pencil or a protractor, ruler and pencil. Illustrations also give ideas on what you could make from these shapes such as medieval houses, Egyptian pyramids, trains, rockets, etc.

Jenkins, Wild. *Make Shapes - Mathematical Models*. Tarquin Publications, Stradbroke, Diss, Norfolk, England. Tel: 037 984 218.
Three pamphlets containing models of polyhedra to trace, cut out and glue together. The finished shapes look fabulous and they have names to match. The polyhedra range from the five platonic solids to more complicated models.

Series No 1   Nineteen simpler models
Tetrahedron, pentagonal prism, pentagonal pyramid, octahedron, truncated tetrahedron, step pyramid, double hexagonal pyramid, truncated square pyramid, cuboctahedron, icosahedron, dodecahedron, rhomboid, truncated octahecron, rhombicuboctahedron, rotatable ring of tetrahedra, tetrahemihexahedron, octahemioctahedron, stella octangula, small stellated dodecahedron.

Series No 2   Eight larger and more complicated models
Faceted cube, icosidodecahedron, cross of octahedron, rhombicosidodecahedron, great stellated dodecahedron, third stellation of the icosahedron, great dodecahedron, compound of five tetrahedra.

Series No 3   Three large and intricate models
Ninth stellation of the icosahedron, great icosahedron, compound of 10 tetrahedra.

Jenkins, Wild. *Mathematical Curiosities*. Tarquin Publications, Stradbroke, Diss, Norfold. Tel: 037 984 218.

Two pamphlets containing directions for mathematical models to draw, cut and glue. Making these models is fun and sparks geometrical thinking. Each book contains nine projects and a minibook.

Book 1
A polyhedra flower, a diabolic frame, a double sided magic square puzzle, three mobius strips, a Klein cube, a double helix, a collecton of shapes of constant width, a family of three hexaflexagons and a set of folding and unfolding cubes.

Book 2
A hexagonal rotating ring, a hypercube, a spherical icosahedron, a rainbow squares puzzle, a cat and mouse tetraflexagon, a square set of nesting pyramids, a magic cube which opens so you can see inside.

Wenninger, M.J. *Polyhedron Models.* Cambridge University Press.
This book contains photographs of all known polyhedra together with templates and suggested methods for constructing each of them.

Cundy, H.M. and Rollett, A.P. *Mathematical Models.* Tarquin Publications, Stradbroke, Diss, Norfolk. Tel: 037 984 218.
Complete nets and instructions are given for all regular (Platonic), Archimedean and stellated polyhedra together with compounds of some of them.

Probability and Statistics

Russell, Susan J, Rebecca Corwin, and Susan Friel. *Used Numbers: Real Data in the Classroom.* Dale Seymour Publications, P.O. Box 10888, Palo Alto, CA 94393-0879.

Patterns and Functions

Davis, Robert B. *Discovery in Mathematics.* Cuisenaire Company of America, 10 Bank Street, White Plains, NY 10606
One of the few resources specifically devoted to patterns and functions for students below high school age. Grades 4-8.

Logic

AIMS. *Primarily Bears.* Aims Educational Foundation, P.O. Box 8120, Fresno, CA 93747-8120.
Thirty-four lesson plans designed around the theme of teddy bears. Six of the lessons contain 20+ problems specifically designated as logic. Integrates math and science with language arts and social studies. Lesson plans with blackline masters. Grades K - 6.

Brisby, Linda Sue et al. *Logic.* Hands on, Inc. 2121 Rebild Drive, Solvang, CA 93463.
Lesson plans arranged by grade level groupings covering attributes, sequencing, relationships,if/then, all/some/none, and/or/not, Venn diagrams, valid/invalid inferences,process of elimination, interpolation, extrapolation and syllogisms. Grades K - 9.

Math Plans

Holden, Linda. *Thinker Tasks, Book 1: Attributes and Logic.* Creative Publications P.O. Box 10328, Palo Alto, California 94303.
Logic problems involving attributes, visual analogies, and money. Includes solutions. Grades 4-6.

Leimbach, Judy. *Primarily Logic.* Dandy Lion Publications, San Luis Obispo, California.
Logic problems - finding relationships and group membership, analogous relationships, deductive reasoning, problem solving and organizing information. Answers included. Grades 2-4.

Post, Beverly and Sandra Eads. *Logic, Anyone?.* Fearon Teacher Aids, a division of Pitman Learning, Inc. 6 Davis Drive, Belmont, California 94002.
All kinds of logic problems to solve - analogies, matrix logic, table logic, circle logic, syllogisms, Venn Diagrams. Includes an answer key. Grades 5 and up.

Reineke, Kathleen, Ed. *The Dell Book of Logic Problems #4.* Dell Readers Service, Box DR, 1540 Broadway, New York, NY 10036.
Compiled by the editors of the Dell Puzzle Magazine, this is the fourth book of logic problems published to meet a seemingly insatiable appetite for using deductive reasoning skills - a la Sherlock Holmes - felt in the general public. The problems are categorized according to their difficulty: easy, medium, hard, challenger. Grades 5 - adult.

Schoenfield, Mark and Jeannette Rosenblatt. *Discovering Logic.* Fearon Teacher Aids, a division of David S. Lake Publishers, 6 Davis Drive, Belmont, California 94002.
Logic problems of classification, sequencing, inference, deduction and creative logic. Answer key. Grades 4 - 6.

Number

Baratta-Lorton, Mary. *Mathematics Their Way.* Addison-Wesley Publishing Co., Inc.
A complete, activity-centered math curriculum for grades K-2.

Brummet, Micaelia Randolph and Linda Holden Charles. *Connections Linking Manipulatives to Mathematics .* Creative Publications, 788 Palomar Avenue, Sunnyvale, CA 94086. Grade 1.

Burke, Donna, Allyn Snider and Paula Symonds. *Math Excursions* and *Box It or Bag It Mathematics,* Cuisenaire Company of America, 10 Bank Street, White Plains, New York 10606-1951. Grades K - 2. Integrated units.

Burns, Marilyn. *Math by All Means, Multiplication: Grade 3*. Cuisenaire Company of America, 10 Bank Street, White Plains, New York 10606-1951.
A five-week unit that provides a model that is an alternative to textbook instruction, grade 3.

Russell, Susan Jo, Rebecca Corwin, and Cornelia Tierney. *Seeing Fractions*. California State Department of Education.
A five-week teaching unit about fractions, grades 4 - 6.

Tyler, Jenny. *Multiplying and Dividing*. Workbook. Usborne Publishing Co., London, England.

All of the strands

Baratta-Lorton, Mary. *Mathematics Their Way*. Addison-Wesley Publishing Co., Inc.
Lessons and lesson ideas at three levels: concept-developing, connecting the concepts to the mathematical symbols and the symblic level. Photographs of the lessons. Grades K-2.

Burns, Marilyn. *About Teaching Mathematics*. Cuisenaire Company of America,10 Bank Street, White Plains, New York 10606-1951.
A rich collection of mathematical lesson ideas and lesson plans in all of the strands. Lessons for whole-class, group and individual work as well as ideas for assessment. Grades K - 8.

Stenmark, Jean Kerr, Thompson, Virginia, and Cossey, Ruth. *Family Math*. Math/ Science Network, Lawrence Hall of Science, University of California, Berkeley, CA 94720.

Periodicals

The Arithmetic Teacher and *Teaching in the Middle School*. 1906 Association Drive, Reston, VA 22091-1593.

Assessment

California State Department of Education. *A Sampler of Mathematics Assessment*. Department of Education, Publication Sales, P.O. Box 271, Cacramento, CA 95802-0271.

California State Department of Education,*California Assessment Program Survey of Basic Skills: Grades 3,6,8 and 12*. Bureau of Publications, Sales Unit, P.O. Box 271, Sacramento, Ca 95802-0271.documents)

Math Plans

Recreational Reading in Math

Pappas, Theoni, *The Magic of Mathematics* and *More Joy of Mathematics*, Wide World Publishing/ Tetra, P.O. Box 496, San Carlos, CA 94070.
From pi to fractals, these books are an eclectic mathematical feast for teachers with a mathematical interest.

Index

A
angles 67, 76, 77, 94
area 133-135, 136-137
arithmetic mean (average) 178-179, 186
attention, signals for 33-35, 371
attributes 248-249

C
California, state of
  Model Curriculum Guide's
    Essential Understandings 36, 42, 44, 47, 49, 52
check for understanding (cfu) 28
circle 67
  circumference 92-93, 131-132
closed/open space 72
computation
  addition 300, 302, 309, 330, 335
  subtraction 306, 315, 330, 335
  multiplication 211, 317, 319, 327, 338
  division 325, 327, 341
congruency and similarity 80
cooperative learning
  three simple rules 31-325
  roles for group members 29-31

D
decimals
  naming and comparing 381-3831
  multiplying 382-386

E
exponents 216-218

F
Fibonacci numbers 196
fractions
  naming 352-354, 355-359, 360-363, 369-372
  comparing 364-367
  addition/subtraction 373-378
  multiplication 384-388
frequencies 186
functions 195-196, 200-203, 208-210, 216-218, 233-237, 242-247

G
geometry
  Euclidean 37, 136-137
  projective 37, 94-96
  topology 37

graphs 90-91, 139, 165, 171-173, 179-181, 185-186,

L
least common multiple (lcm) 321

M
make your own
  cube 86-89
  dodecahedron 86-89
  fraction bars 369
  hundreds chart 415
  icosahedron 86-89
  spinner 174, 406
  tangram 101
  tesselation 103-106
  tetrahedron 86-89
mod arithmetic 211-215
money 341-351

N
numbers
  composite/prime 319-320
  even/odd 197
  Fibonacci 196
  square 227-229

P
percent 395-403
perimeter 224-226
pi 92-93, 131-132
Platonic solids 86-89
polyhedrons 86-89
prime numbers 319-320

R
rates 389-391
ratio 392-394

S
symmetry
  line 69
  rotational 83

T
triangle inequality 97

V
van Hiele levels in geometry 38-40

W
Waldorf School 40-41